FAA-STD-019d
August 9, 2002

DEPARTMENT OF TRANSPORTATION
FEDERAL AVIATION ADMINISTRATION STANDARD

LIGHTNING AND SURGE PROTECTION, GROUNDING, BONDING AND SHIELDING REQUIREMENTS FOR FACILITIES AND ELECTRONIC EQUIPMENT

FOREWORD

This document mandates standard configurations and procedures for lightning protection, surge and transient protection, grounding, bonding and shielding. All construction of Federal Aviation Administration (FAA) operational facilities and the electronic equipment installed therein shall conform to this standard. This document defines minimum requirements for all FAA facilities. The specific needs of any facility may exceed these minimum requirements. These needs may be influenced by the equipment to be installed at the site, the configuration of the structures and location of the equipment, and by the physical environment present at the location.

The interface between contractor owned equipment or electronic equipment not used for operational purposes (administrative local area network (LAN), administrative telephone, etc.) and the operational facility shall be in accordance with this document. This standard applies to new construction and modifications to existing facilities, particularly when required for the installation of electronic equipment.

The use of shall in this document indicates mandatory compliance. However, in certain cases it may not be technically feasible to implement certain requirements. In these cases, a National Airspace System (NAS) Change Proposal (NCP) must be submitted with adequate justification and technical documentation and approved by the NAS Configuration Control Board (CCB) before a deviation is permitted.

This standard contains six sections. Section 1 details the scope and purpose of the standard. Section 2 lists reference documents. Section 3 gives requirements for surge and transient protection, lightning protection, the EES (EES), the electronic multipoint ground system (MPG), the electronic single point ground system (SPG), and National Electrical Code (NEC) compliance. Section 4 provides quality assurance requirements. Section 5, "Preparation for Delivery", does not apply to this document. Section 6 contains notes, definitions, acronyms, and abbreviations.

TABLE OF CONTENTS

1. SCOPE 1
   1.1 Scope 1
   1.2 Purpose 1
2. APPLICABLE DOCUMENTS 3
   2.1 Government Documents 3
   2.2 Non-Government Documents 4
3. REQUIREMENTS 7
   3.1 Surge and Transient Protection Requirements 7
      3.1.1 General 7
      3.1.2 Existing Electronic Equipment Designs 7
   3.2 External Lines and Cables 8
      3.2.1 Fiber Optic Lines 8
      3.2.2 Balanced Pair Lines 8
      3.2.3 Ferrous Conduit 8
      3.2.4 Buried Guard Wires 9
      3.2.5 Armored Cable (Direct Earth Burial (DEB) Type) 9
      3.2.6 Existing Metallic Conduit, Conductors and Cables 9
   3.3 Interior Lines and Cables 9
   3.4 Electronic Equipment Transient Susceptibility Levels 10
   3.5 Conducted Power Line Surges 10
      3.5.1 Surge Levels 10
      3.5.2 Facility AC Surge Protective Device 11
      3.5.3 Surge Protective Devices for Distribution and Branch Panels 13
      3.5.4 Electronic Equipment Power Entrance 13
      3.5.5 DC Power Supply Transient Suppression 14
   3.6 Conducted Signal, Data and Control Line Transients 15
      3.6.1 Transient Levels 15
      3.6.2 Protection Design 16
      3.6.3 Characteristics 16
      3.6.4 Installation of Facility Level Transient Protection 17
      3.6.5 Installation of Suppression Components at Electronic Equipment 18
      3.6.6 Externally Mounted Electronic Equipment 18
      3.6.7 Axial Cables 18
      3.6.8 Fiber Optic Cable 19
   3.7 Lightning Protection System Requirements 20
      3.7.1 General 20
      3.7.2 Materials 20
      3.7.3 Main Conductors 20
      3.7.4 Hardware 20
      3.7.5 Guards 21
      3.7.6 Bonds 21
      3.7.7 Conductor Routing 22
      3.7.8 Down Conductor Terminations 22

3.7.9	Buildings	22
3.7.10	Antenna Towers	24
3.7.11	Fences	27
3.7.12	Airport Traffic Control Towers (ATCT)	29

3.8 Earth Electrode System (EES) Requirements 32
    3.8.1 General 32
    3.8.2 Site Survey 32
    3.8.3 Design 32
    3.8.4 Configuration 35
    3.8.5 Ground rods 35
    3.8.6 Interconnections 36
    3.8.7 Access Well 36

3.9 Main Ground Plate 36

3.10 Electronic Multipoint Ground System Requirements 37
    3.10.1 General 37
    3.10.2 Facilities Requiring a Signal Reference Structure (SRS) 37
    3.10.3 Ground Plates, Cables and Protection 39
    3.10.4 Building Structural Steel 41
    3.10.5 Interior Metallic Piping Systems 42
    3.10.6 Electrical Supporting Structures 42
    3.10.7 Secure Facilities 43
    3.10.8 Multipoint Grounding of Electronic Equipment 43
    3.10.9 Electronic Equipment Containing both Low and High Frequency Circuits 44

3.11 Electronic Single Point Ground System Requirements 44
    3.11.1 General 44
    3.11.2 Ground Plates 44
    3.11.3 Isolation 45
    3.11.4 Resistance 45
    3.11.5 Ground Cable Size 45
    3.11.6 Labeling 46
    3.11.7 Equipment Requiring Electronic Single Point Grounds 46

3.12 National Electrical Code (NEC) Grounding Compliance 49
    3.12.1 General 49
    3.12.2 Grounding Electrode Conductors 50
    3.12.3 Equipment Grounding Conductors 50
    3.12.4 Color Coding of Conductors 51
    3.12.5 Conductor Routing 52
    3.12.6 Non-Current-Carrying Metal Equipment Enclosures 52

3.13 Other Grounding Requirements 53
    3.13.1 Electronic Cabinet, Rack, and Case Grounding 53
    3.13.2 Receptacles 54
    3.13.3 Equipment Power Isolation Requirements 54
    3.13.4 Portable Equipment 54
    3.13.5 Fault Protection 54
    3.13.6 AC Power Filters 55

3.14 Bonding Requirements 55
    3.14.1 Resistance 55
    3.14.2 Methods 55
    3.14.3 Bonding Straps and Jumpers 59
    3.14.4 Fasteners 60
    3.14.5 Temporary Bonds 60

	3.14.6	Inaccessible Locations	60
	3.14.7	Coupling of Dissimilar Metals	61
	3.14.8	Surface Preparation	61
	3.14.9	Bond Protection	62
	3.14.10	Bonding Across Shock Mounts	63
	3.14.11	Enclosure Bonding	63
	3.14.12	Subassemblies	63
	3.14.13	Equipment	63
	3.14.14	Connector Mounting	63
	3.14.15	Shield Terminations	64
	3.14.16	RF Gaskets	64

3.15 Shielding Requirements 65
    3.15.1 Design 65
    3.15.2 Facility Shielding 65
    3.15.3 Conductor and Cable Shielding 65
    3.15.4 Space Separation 67
    3.15.5 Electromagnetic Environment Control 68

3.16 Electrostatic Discharge (ESD) Minimization, Control and Prevention Requirements 70
    3.16.1 ESD Sensitivity Classification 70
    3.16.2 ESD Protection Requirements 70
    3.16.3 Circuit and Equipment Design 71
    3.16.4 Classification of Materials 71
    3.16.5 Protection of ESD Susceptible and Sensitive Items 72
    3.16.6 Hard and Soft Grounds 76
    3.16.7 ESD Control Floor and Coverings 76
    3.16.8 ESD Protective Worksurfaces 77
    3.16.9 Static Dissipative Coatings 79

4. QUALITY ASSURANCE PROVISIONS 80

4.1 Electromagnetic Compatibility and Quality Assurance 80
    4.1.1 General 80

4.2 Requirements 80

4.3 Approval 80

5. PREPARATION FOR DELIVERY 82

6. NOTES 84

6.1 Definitions 84
    6.1.1 Access Well 84
    6.1.2 Air Terminal 84
    6.1.3 Armored Cable 84
    6.1.4 Arrester 84
    6.1.5 Bond 84
    6.1.6 Bond, Direct 84
    6.1.7 Bond, Indirect 84
    6.1.8 Bonding 84
    6.1.9 Bonding Jumper 84
    6.1.10 Branch Circuit 84
    6.1.11 Brazing 85
    6.1.12 Building 85
    6.1.13 Bulkhead Plate 85
    6.1.14 Cabinet 85

6.1.15	Cable	85
6.1.16	Case	85
6.1.17	Chassis	85
6.1.18	Clamp Voltage	85
6.1.19	Conductor	86
6.1.20	Crowbar	86
6.1.21	Current Issue	86
6.1.22	Earth Electrode System (EES)	86
6.1.23	Electromagnetic Interference (EMI)	86
6.1.24	Electronic Multipoint Ground System	87
6.1.25	Electronic Single Point Ground System	87
6.1.26	Equipment Areas	87
6.1.27	Equipment Grounding Conductor	87
6.1.28	Equipment	87
6.1.29	Facility Ground System	87
6.1.30	Faraday Cage	87
6.1.31	Feeder	87
6.1.32	Fitting, High Compression	87
6.1.33	Ground	88
6.1.34	Grounded	88
6.1.35	Grounded Conductor	88
6.1.36	Grounded, Effectively	88
6.1.37	Grounding Conductor	88
6.1.38	Grounding Electrode	88
6.1.39	Grounding Electrode Conductor	88
6.1.40	High frequency	88
6.1.41	Issue, Current	88
6.1.42	Landline	88
6.1.43	Line Replaceable Unit	89
6.1.44	Low Frequency	89
6.1.45	National Electrical Code (NEC) (NFPA-70)	89
6.1.46	Operational Areas	89
6.1.47	OPI	89
6.1.48	Overshoot Voltage	89
6.1.49	Pressure Connector	89
6.1.50	Rack	89
6.1.51	Reference Plane or Point, Electronic Signal (Signal Ground)	89
6.1.52	Reverse Standoff or Maximum Continuous Operating Voltage (MCOV)	89
6.1.53	Shield	89
6.1.54	Structure	90
6.1.55	Surge	90
6.1.56	Susceptibility Level	90
6.1.57	Transient	90
6.1.58	Transient Suppressor	90
6.1.59	Turn-on Voltage	90
6.1.60	Zone of Protection	90

6.2 Acronyms and Abbreviations 91

6.3 Guidelines 92

List of Figures

Figure 1. Lightning Protection for Radomes and Radar Antenna Platforms 25
Figure 2. Grounding a Fence 28
Figure 3. Airport Traffic Control Tower Levels 30
Figure 4. Ground Plate Detail 34
Figure 5. Grounding Trench Detail 35
Figure 6. Facility Grounding System 38
Figure 7. Multipoint Ground Cable Size Determination 40
Figure 8. Electronic Single Point Ground System Installation 47
Figure 9. Single Point Electronic Ground Bus Bar Installation in Rack or Cabinet 49
Figure 10. Bonding of Conduit and Grounding Conductor 53
Figure 11. Order of Assembly for Bolted Connections 58
Figure 12. Bonding of Connectors to Mounting Surface 64
Figure 13. Grounding of Overall Cable Shields to Connectors and Penetrating Walls 67
Figure 14. Grounding of Overall Cable Shield to Terminal Strip 68

List of Tables

Table I. Surge Levels for 120/208V, 120/240 and 277/480V Service Lines 11
Table II. Conducted Landline Transient Levels 15
Table III. Size of Electronic Multipoint Ground Interconnecting Conductors 41
Table IV. Minimum Number of Twists for Power Conductors 53
Table V. Acceptable Couplings Between Dissimilar Metals 57
Table VI. Torque Requirements for Bolted Bonds 58
Table VII. Minimum Separation Distance Between Signal and Power Cables 68

1. SCOPE

1.1 Scope

This document mandates standard lightning protection, transient protection, electrostatic discharge (ESD), grounding, bonding and shielding configurations and procedures for new facilities, facility modifications, facility upgrades, new equipment installations, and new electronic equipment used in the National Airspace Systems (NAS). It provides requirements for the design, construction, modification or evaluation of facilities and equipment. This document does not apply to existing facilities unless the facility is undergoing upgrade or receiving new electronic equipment. This version of the document applies to equipment and facilities and procurement initiated after the effective date of this document. However, if the procurement contract is not in accordance with the version of FAA-STD-019 in effect on the initiation date of the procurement, then the procurement shall be required to meet the version of this standard in effect when the noncompliance is noted.

1.2 Purpose

The requirements of this standard provide a systematic approach to minimize electrical hazards to personnel, electromagnetic interference and damage to facilities and electronic equipment from lightning, transients, ESD, and power faults.

2. APPLICABLE DOCUMENTS

2.1 Government Documents

The current issues of the following documents form a part of this standard and are applicable to the extent specified elsewhere in this document. If conflicts occur between these documents and the contents of this standard, the contents of this standard provide the superseding requirements.

<u>Federal Specifications</u>

P-D-680   Dry Cleaning Solvent

(Information required to obtain copies of federal specifications is available from General Services Administration offices in Atlanta, Auburn WA; Boston, Chicago, Denver, Forth Worth, Kansas City MO, Los Angeles, New Orleans, New York, San Francisco, and Washington DC)

<u>FAA Specifications</u>

FAA-C-1217	Electrical Work, Interior
FAA-G-2100	Electronic Equipment, General Requirements

<u>FAA Standards</u>

FAA-STD-012   Paint Systems for Equipment

<u>FAA Orders</u>

Order 6950.19	Practices and Procedures for Lightning Protection, Grounding, Bonding and Shielding Implementation
Order 6950.20	Fundamental Considerations of Lightning Protection, Grounding, Bonding and Shielding

(Copies of these specifications, standards, orders, and other applicable FAA documents may be obtained from the Contracting Officer issuing the invitation-for-bids or request-for-proposals. Requests should fully identify material desired, i.e. specification, standard, amendment, drawing numbers and dates. Requests should cite the invitation-for-bids, request-for-proposals, the contract involved, or other use to be made of the requested material.)

<u>Military Documents</u>

MIL-HDBK-253	Guidance for the Design and Test of Systems Protected Against the Effects of Electromagnetic Energy
DOD/MIL-HDBK-263	Electrostatic Discharge Control Handbook
DOD-STD-1686	Electrostatic Discharge Control Program for Protection of Electrical and Electronic Parts, Assemblies and Equipment (Excluding Electrically Initiated Explosive Devices)

MIL-HDBK-237	Electromagnetic Compatibility Management Guide for Platforms, Systems and Equipment
MIL-HDBK-419	Grounding, Bonding, and Shielding for Electronic Equipment and Facilities
MIL-PRF-87893	Performance Specification, Workstations, Electrostatic Discharge Control
MIL-STD-461	The Control of Electromagnetic Interference Emissions and Susceptibility
MIL-STD-889	Dissimilar Metals
MIL-STD-1686	Electrostatic Discharge Control Program for Protection of Electrical and Electronic Parts, Assemblies, and Equipment Excluding electrically Initiated Explosive Devices
MIL-W-87893	Workstation, Electrostatic Discharge Control
NACSIM 5203	Guidelines for Facility Design and Red/Black Installation (U) (Confidential Document)

(Single copies of Military specifications, standards, and handbooks may be requested by mail or telephone from Document Automation and Production Service Customer Service Standardization Documents Order Desk 700 Robbins Avenue, Building 4D Philadelphia, PA 19111-5094 or via www.dodssp.daps.mil/dodssp.htm. Not more than five items may be ordered on a single request; the Invitation for Bid or Contract Number should be cited where applicable. Only latest revisions (complete with latest amendments) are available; slash sheets must be individually requested. Request all items by document number

2.2 Non-Government Documents

The current issues of the following documents form a part of this standard and are applicable to the extent specified herein. If conflicts occur between these documents and the contents of this standard, the contents of this standard provide the superseding requirements.

Electronic Industries Alliance (EIA)

EIA Standard EIA-625 www.eia.org	Requirements for Handling Electrostatic-Discharge-Sensitive (ESDS) Devices

Requests for copies of EIA Standards should be addressed to Electronic Industries Alliance, Corporate Engineering Department, 2500 Wilson Blvd, Arlington, VA 22201 or telephone 703 907-7500.

National Fire Protection Association (NFPA)

NFPA 70	National Electrical Code (NEC)
NFPA 77	Static Electricity
NFPA 780	Standard for the Installation of Lightning Protection Systems

(Requests for copies of NFPA documents should be addressed to the National Fire Protection Association, One Batterymarch Park, Quincy MA 02269. www.nfpa.org)

Underwriters Laboratories, Inc. (UL)

UL 96	Lightning Protection Components
UL 96A	Installation Requirements for Lightning Protection Systems
UL 779 (ANSI-A148.1)	Electrically Conductive Floorings
UL 1449	Transient Voltage Surge Suppressors

(Requests for copies of UL documents should be addressed to Global Engineering Documents, 1500 Inverness Way East Englewood, CO 80112   Telephone 303 397-7945, 800 854-7179 www.ul.com)

Institute of Electrical and Electronic Engineers (IEEE)

ANSI/IEEE C62.41	Recommended Practice on Surge Voltages in Low Voltage AC Power Circuits
ANSI/IEEE C62.45	IEEE Guide on Surge Testing for Equipment Connected to Low-Voltage AC Power Circuits
ANSI/IEEE 1100	Recommended Practice for Powering and Grounding Sensitive Electronic Equipment (Emerald Book)

(Requests for copies of IEEE documents should be addressed to Institute of Electrical and Electronic Engineers, 445 Hoes Lane, P.O. Box 1331, Piscataway, NJ 08855-9916. www.ieee.org)

Electrostatic Discharge (ESD) Association Documents

ESD-ADV 2.0	Electrostatic Discharge (ESD) Control Handbook
ESD-ADV 53.1	ESD Protective Workstations
ANSI-EOS/ESD-S 4.1	ESD Protective Work Surfaces Resistance Characterization
ANSI/ESD-S 7.1	Resistive Characterization of Materials - Floor Materials
ESD-S 11.11	Surface Resistance Measurement of Static Dissipative Planar Materials
ESD S20.20	Development of an Electrostatic Discharge Control Program
ESD-STM 4.1	Work Surfaces - Resistance Measurements
ESD-STM 5.1	Sensitivity Testing, Human Body Model, Component Level
ESD-STM 12.1	Seating - Resistive Measurement

(Requests for copies of ESD Association documents should be addressed to the ESD Association, 7900 Turin Road, Bldg 3, Suite 2, Rome, NY 13440-2069. www.esda.org telephone 315 339-6937)

3. REQUIREMENTS

This section provides requirements that are established to insure the proper operation of FAA facilities. The use of the term facilities in this document may differ from the manner in which it is frequently used in other FAA documents. For the purposes of this document, a facility is an area of collocated equipment. For example, the cab and electronic/electrical equipment located on the junction and subjunction levels of an Airport Traffic Control Tower (ATCT) are a single facility. An ATCT with a base building containing electronic equipment is an example of two facilities located at the same site. Other examples of 2 or more facilities (in the sense of the term in this document) include the ARSR-4 (the tower and base building are separate facilities) and Air Route Traffic Control Centers (ARTCC) with multiple buildings that must be treated as separate facilities. An example of a single facility (for purposes of this document) is a Remote Controlled Air to Ground (RCAG) collocated in a VHF Omni-directional Range (VOR) building. Contact the Office of Primary Interest (OPI) of this document for specific guidance on new facilities/systems.

3.1 Surge and Transient Protection Requirements

3.1.1 General

Lines, cables, and facility electronic equipment shall be protected against damaging surges on alternating current (AC) power lines and transients on electronic landlines from the effects of lightning. Ferrous conduit shall be used to shield external lines and cables to minimize inductive coupling of transients from lightning discharges. Guard wires shall be used to protect buried cables from direct lightning strikes. Fiber optic lines and balanced metallic lines shall also be used when feasible. Electrical and electronic equipment shall be protected against conducted and radiated surges and transients from all power, signal, control and/or status lines. Integrated circuits, transistors, diodes, solid-state voltage regulators, capacitors, miniature relays, miniature switches, and miniature transformers, etc. are quite susceptible to damage and operational upset caused by transients. Transient suppression shall be provided at the entrance of lines and cables to facility structures and electronic equipment enclosures as required in paragraphs 3.6 and 3.7 and their associated subparagraphs to protect electronic equipment from conducted transients. A surge protective device (SPD), capable of shunting the energy represented by the surge levels in paragraph 3.5.1, shall be installed at the AC service entrance to the facility. Implementation guidelines are contained in FAA Orders 6950.19 and 6950.20. All surge protective devices shall meet or exceed the current version of UL 1449. Documentation, including a schematic diagram of the SPD circuitry and complete installation instructions shall be provided for all SPDs installed at FAA facilities. The manufacturer's data sheet for the SPD shall be filed in the Facility Reference Data File (FRDF).

3.1.2 Existing Electronic Equipment Designs

For existing electronic equipment designs, the equipment level transient suppression required by this document, may be either internal or external to the electronic equipment using proper grounding, bonding, and shielding procedures.

3.2 External Lines and Cables

3.2.1 Fiber Optic Lines

Fiber optic lines are not inherently susceptible to electromagnetic interference or the induction fields produced by lightning. Fiber optic lines should replace metallic lines when economically and technically feasible. Ferrous conduit shielding is not required for fiber optic lines. Suppression components are not required for fiber optic lines. Where metallic or electrically conductive sheaths or strength members are present they shall be grounded to the multipoint ground system at each end. The fiber optic modules shall have 90db of attenuation against all sources of electromagnetic interference (EMI).

3.2.2 Balanced Pair Lines

When possible, signal and control circuits routed external to equipment shall be designed as balanced, two conductor, shielded circuits.

3.2.3 Ferrous Conduit

In this standard, ferrous conduit is defined as rigid galvanized steel (RGS) or ferrous rigid metal conduit. Buried power lines and all buried electronic (signal, control, communication, RMM, etc.) lines, conductors and cables to the facility except axial type lines shall be enclosed in galvanized steel conduit. However, where the length of the run exceeds 300 ft., the above referenced lines, conductors and cables into the facility shall be either in armored cable (with a ferrous armor) as defined in paragraph 3.2.5 or enclosed in continuous RGS. Buried axial type lines shall have a ferrous metallic armor or be run in intermediate metal conduit. All power, signal and control lines, conductors and cables, both overhead and buried, shall enter the facility through a minimum of 10 feet of ferrous conduit. The conduit may be routed via other than a direct route to achieve this 10 foot requirement. This conduit, if buried, shall extend 5 feet beyond the EES. The buried end(s) of each of these conduits shall be bonded to the EES with a bare copper conductor, #2 American Wire Gauge (AWG), minimum. Conduit joints and fittings shall be electrically continuous with bonding resistance of 5 milliohms or less between joined parts. Conduit enclosing AC power service entrance conductors shall be bonded to the distribution transformer case and to the service entrance cabinet. Conduit enclosing signal, control, status, power, or other conductors to electronic equipment shall be terminated using conductive fittings to their respective junction boxes, equipment cabinets, enclosures, or other grounded metal structures.

3.2.3.1 Ferrous Conduit Penetration of Facility

(1) At each location where above ground conduits first penetrate a shelter or building's exterior wall, bonding connections shall be made. The conduit shall be bonded either to a bulkhead connector plate that is connected directly to the EES in accordance with the requirements in paragraph 3.6.7.2 or directly from the conduit to the EES.

(2) If the connections in 3.2.3.1(1) are not feasible, bonding connections shall be made from the conduit to the electronic multipoint ground plate. Conduits shall be connected to an electronic multipoint ground plate with minimum 2 AWG stranded copper conductor using exothermic welds or UL listed pressure connectors.

3.2.4 Buried Guard Wires

Buried lines including armored cable, not completely enclosed in ferrous conduit, shall be protected by a bare #1/0 AWG copper guard wire. The guard wire shall be embedded in the soil, a minimum of 10 in. (25cm) directly above and parallel to the lines or cables being protected. When the width of the cable run or duct does not exceed 3 ft (90cm), one guard wire, centered over the cable run or duct, provides adequate protection. When the cable run or duct is more than 3ft (90cm.) in width, 2 guard wires shall be installed. The guard wires shall be spaced at least 12 in.(30 cm.) apart and be not less than 12 in. (30 cm.) nor more than 18 in. (45 cm.) inside the outermost wires or the edges of the duct. The guard wire shall be bonded to the EES at each end and to ground rods at approximately 90ft intervals using exothermic welds. Where cables are run parallel to the edge of a runway an additional guard wire located between the runway edge and the cable run has been shown to provide significant reduction in lightning related incidents. The spacing between ground rods must vary by 10 – 20% to prevent resonance. Install the ground rods at approximately 6 feet (2 m) to either side of the trench.

3.2.5 Armored Cable (Direct Earth Burial (DEB) Type)

Armored cable (with a ferrous armor), when permitted according to the requirements of paragraph 3.2.3, shall be installed with a guard wire in accordance with paragraph 3.2.4. Armored cable shall be bonded to the EES prior to the point of entry into the facility with a minimum #2 AWG bare copper conductor. Where this is not feasible, armor shall be bonded to the multipoint ground (MPG). When none of the above are available, armor shall be grounded by bonding to the ground bus at the service disconnecting means. If armor is continued to the electronic equipment, it shall be bonded to the multipoint ground system of the electronic equipment unless the equipment is required to be isolated. All bonds shall be less than 5 milliohms between joined parts.

3.2.6 Existing Metallic Conduit, Conductors and Cables.

All conduit and cables in NAS operational facilities are subject to currents induced by nearby lightning strikes. These induced effects can adversely affect the operation of sensitive electronic equipment. All unused conduits, conductors and cables shall be removed, unless the removal is disapproved by the Airway Facilities (AF) manager. The AF manager shall be consulted to validate the decision to remove any metallic conduit, conductors or cables prior to acting. If they are to remain, the following actions shall be accomplished to minimize the voltage differential between ends.
- All metallic conduits shall be grounded at both ends.
- All unused conductors shall be grounded at both ends.
- In unused cables, ground all conductors, including shields, at both ends. This grounding may be accomplished through a gas tube at one end to minimize hum.

3.3 Interior Lines and Cables

All permanently installed single conductors, cables and wiring exceeding 6 ft. in length shall be enclosed in steel conduit, shielded cable trays, or shielded wireways that are connected to the electronic multipoint ground system as specified in paragraphs 3.10.6.1 and (e). Backbone type and other cable trays that do not provide shielding from EMI and adequate bonding surface area are strictly prohibited.

3.4 Electronic Equipment Transient Susceptibility Levels

Electronic equipment, such as radars, navaids, transmitters, supplied as part of the facility, shall be provided with transient protection that shall reduce surges and transients of 2.5 times the normal operating voltage or 600 volts whichever is larger, to below the equipment susceptibility level. Electronic equipment not supplied as a part of the facility would include items such as administrative computers or other similar equipment not required for the facility to perform its operational function. The equipment susceptibility level is defined as the transient level on the signal, control or data line that may cause damage, degradation, or upset to electronic circuitry connected to the line. Protection for these levels is in addition to the facility protection levels specified in paragraphs 3.5 and 3.6. The electronic equipment manufacturer shall perform tests to determine the voltage, current, or energy levels that will cause immediate damage to components, shorten its operating life, or cause operational upset. These tests shall consider all electrical and electronic equipment components exposed to the effects of surges or transients. The combined facility and equipment entrance protection shall be coordinated to limit transients at the equipment to below the equipment susceptibility level. Requirements of this paragraph shall be included in the comprehensive control and test plans outlined in paragraph 4.2. In all cases the following characteristics shall be evaluated.

(a) <u>Component damage threshold</u>. The damage threshold is the transient level that renders the component nonfunctional or operationally deficient. For solid-state components, voltage is usually the relevant parameter.

(b) <u>Component degradation level</u>. The component degradation level is the transient voltage or energy level that shortens the useful life of the component.

(c) <u>Operational upset level</u>. The operational upset level is the transient voltage or energy level that causes an unacceptable change in operating characteristics for longer than 10 milliseconds for analog equipment or a change of logic state for digital equipment.

3.5 Conducted Power Line Surges

A surge protective device (SPD) shall be provided at the service disconnecting means. This entrance arrester may be located on the line side of the disconnecting means if properly rated for the application. Additional SPDs shall be provided at power line entrances to operational electronic equipment. SPDs at the service disconnecting means, distribution and branch panel boards as well as transient suppression provided at electronic equipment power line entrances shall be coordinated in accordance with the guidance provided in paragraphs 3.6.2.1(a) and 3.6.4.1 (a).

3.5.1 Surge Levels

Surge levels and number of occurrences for selection or design of facility AC arresters are given by Table I. Table I defines line-to-ground, line-to-neutral, neutral to ground, and line-to-line surge currents, and number of occurrences for low voltage AC services. In these tables, the 8-by-20 μs wave form defines a transient reaching peak value in 8 μs and decays to 50 percent of peak value 20 μs after inception. These devices shall be able to tolerate surges of shorter duration without malfunction.

Table I. Surge Levels for 120/208V, 120/240 and 277/480V Service Lines

Surge Current Amplitude 8-by-20 Microsecond Waveform *4-by-10 Microsecond	Number of Surges (Lifetime)
10 kA	1,500
20 kA	700
30 kA	375
40 kA	50
50 kA	8
60 kA	6
70 kA	4
100 kA*	2
180-240kA*	1

3.5.2 Facility AC Surge Protective Device

A facility AC surge protective device (SPD) shall be installed, on the load side of the facility service disconnecting means. However, if it is specifically listed/rated for line side applications, the SPD may be installed on the line side of the facility service disconnect. Note: SPDs listed under UL standard 1449 are not intended for use on the line side of the disconnecting means. When installed on the line side of the facility service disconnecting means, the SPD shall have a 200,000 AIC rated overcurrent protective device in case of a catastrophic failure. The SPD may be a combination of solid state circuits, varistors, or other devices and shall meet the requirements provided in this paragraph and its subparagraphs. For services with a neutral, SPD elements shall be connected line-to-ground (L-G), line-to-neutral (L-N), and neutral-to-ground (N-G). For services without a neutral, SPD elements shall be connected line-to-ground (L-G), and line-to-line (L-L). Lightning arresters (high voltage) shall also be installed on the primary side of FAA owned distribution transformers. These arresters and SPDs shall be approved by the office of primary interest (OPI) of this document.

3.5.2.1 Characteristics

Minimum functional and operational characteristics of facility SPDs for installation at service disconnecting means shall be as follows:

(a) <u>Maximum continuous operating voltage (MCOV)</u>. The maximum designated root-mean-square (rms) value of power frequency voltage that may be applied continuously between the terminals of the arrestor under all non-transient conditions.

(b) <u>Leakage current</u>. The DC leakage current shall be less than 1mA for voltages at or below 1.414 x MCOV VDC.

(c) <u>Clamp (discharge) voltage</u>. The maximum clamp voltage for a SPD when passing up to a 70kA 8/20 microsecond wave shape shall not exceed the value $V_{cl} < (2.7 \times V_{rms}) + (8 \times I_s)$

Where

V_{cl} = The maximum clamping voltage

V_{rms} = The root mean square value of the nominal system voltage

I_s = The short circuit surge current of the surge in kiloamperes.

(d) <u>Overshoot voltage</u>. Overshoot voltage shall not exceed 2 times the SPD clamp voltage for more than 10 nanoseconds. Overshoot voltage is the surge voltage level that appears across the SPD terminals before the device turns on and clamps the surge to the specified voltage level.

(e) <u>Self-restoring capability</u>. The SPD shall automatically return to an off state after surge dissipation when line voltage returns to normal.

(f) <u>Operating lifetime</u>. The SPD shall safely dissipate the number and amplitude of surges listed in Table I. Clamp (discharge) voltage shall not change more than 10 percent over the operating life of the arrester.

(g) <u>Fusing and Indicator Lamps</u>. If the input to an SPD is internally fused this fusing shall not increase the clamp voltage of the SPD and shall pass the surge current levels given by Table I up to the 70kA level without opening. Two indicator lamps per phase on the SPD enclosure cover shall visually indicate that fuse(s) have opened.

3.5.2.2 Packaging

All components comprising a surge protective device shall be packaged in a single National Electrical Manufacturers Association (NEMA) type 12 steel enclosure for indoor use only, or a type 4 waterproof, steel enclosure for indoor or outdoor use. SPDs may be enclosed within panelboards or switchgear enclosures, if UL listed for such installation. Either studs or connectors for #4 AWG or larger conductors shall be provided for all input and output connections. The SPD elements, when not connected to the phase, neutral and ground conductors, shall be electrically isolated from the enclosure to a minimum of 10 megohms resistance. The enclosure door shall be hinged and electrically bonded to the enclosure. Hinges shall not be used to provide electrical bonding. Fuses, lights, fuse wires, and arrester elements or components shall be readily accessible for inspection and replacement.

3.5.2.3 Installation

The SPD shall be installed as close as possible (within 12 in.) of the facility service disconnecting means. Wiring connections may be other than the gauge specified herein if recommended by the manufacturer. Connections shall be made with UL listed pressure connectors.

(a) <u>Phase connections</u>. Phase lugs of the SPD shall be connected to corresponding phase terminals of the service disconnecting means with insulated #4 AWG (minimum) copper cable. Connections shall be as short and direct as possible without loops, sharp bends or kinks.

(b) <u>SPD ground and neutral connections.</u> The ground connection for the SPD elements shall be routed as directly as possible, with no loops, sharp bends or kinks from the SPD element output terminal to the ground bus in the service disconnecting means. In a grounded system, the neutral connection for the SPD elements shall be routed in a similar manner as above to the neutral bus of the service disconnecting means. These ground and neutral connections shall be a #4 AWG (minimum) copper cable, insulated and color coded in accordance with NEC. The element terminals shall be electrically insulated from the SPD enclosure.

(c) <u>Equipment grounding conductor</u>. The SPD enclosure shall be connected to the ground bus in the service disconnecting means enclosure with a minimum of #4 AWG copper wire. The wire shall have green insulation.

(d) <u>Conduit sealing.</u> The conduit or conduit nipple connecting the SPD enclosure to the service disconnecting means (SDM) enclosure shall be sealed with duct seal or other nonflammable medium to prevent soot from entering the SDM enclosure in the event of SPD failure.

3.5.3 Surge Protective Devices for Distribution and Branch Panels

Surge protective devices shall be installed on all critical and essential panels providing service to NAS operational equipment or supplying exterior circuits. Examples of exterior circuits include obstruction lights, exterior convenience outlets, guard shacks, electric gates and feeds to other facilities. SPDs shall be selected in accordance with the guidance provided in IEEE C62.41 and meet the requirements of UL1449-II. Devices for panels serving exterior circuits shall be tested for a level C3 application per IEEE C62.41. These devices shall be installed as close as possible to the panel they serve and in accordance with the manufacturers instructions. The conduit or conduit nipple connecting the SPD enclosure to the panel enclosure shall be sealed with duct seal or other nonflammable medium to prevent soot from entering the enclosure in the event of SPD failure. The use of potting material in SPDs is strictly prohibited. All SPD components must be accessible for inspection by qualified FAA personnel. The MCOV for SPDs located at branch and distribution panels shall be equal to or greater than the MCOV of those located at the facility service.

3.5.4 Electronic Equipment Power Entrance

Surge protective devices, components or circuits for protection of electronic equipment power lines shall be provided as an integral part of all electronic equipment. These devices shall be positioned at the AC power conductor entrance to electronic equipment provided as part of the facility. Transient protection shall be provided on all combinations of L-L, L-G, L-N and N-G. The MCOV for SPDs located at the equipment shall be equal to or greater than the MCOV of those located at branch and distribution panels. SPDs at equipment shall provide a clamping level less than the equipment susceptibility level as defined in paragraph 3.4. The OPI for this document shall approve the design and selection of these devices. Electronic equipment that is to be installed outside of facilities shall also require protection to the level supplied for the facility.

3.5.4.1 Characteristics

The basic characteristics of surge suppression components or circuits for equipment power lines entrances shall be as follows:

(a) <u>Maximum Continuous Operating Voltage (MCOV)</u>. MCOV of the arrester shall be ≥130 percent of the nominal line voltage.

(b) <u>Turn on voltage</u>. Turn on voltage, discharge (clamp) voltage, and the amplitude and time duration of any overshoot voltage shall be less than the equipment susceptibility level.

(c) <u>Leakage current</u>. The DC leakage current shall be less than 1mA for voltages at or below 1.414 x MCOV VDC. .

(d) <u>Self-restoring capability</u>. The surge suppressors shall automatically restore to an off state when surge voltage falls below the turn on voltage for the suppressor.

(e) <u>Operating lifetime</u>. Clamp voltage shall not change more than 10 percent over the operating lifetime. When not located within a facility protected by the SPD required in paragraph 3.5.2 the electronic equipment surge suppression shall be capable of safely dissipating the number and amplitude of surges specified in Table I.

3.5.4.2 Packaging

Suppression components shall be housed in a separate, shielded, compartmentalized enclosure as an integral part of the electronic equipment design. Bulkhead-mounted feed through capacitors or equivalent shall be used as necessary to prevent high frequency transient energy from coupling to electronic equipment circuits. Suppression components shall be grounded to the electronic equipment case as directly as feasible.

3.5.5 DC Power Supply Transient Suppression

Power supplies that use 60 hertz (Hz) power and furnish DC operating voltages to solid state equipment used in direct support of the NAS, shall have transient suppression components from each output of the power supply to the equipment chassis. During conduction of transients, the suppressor shall not decrease rectifier output voltage below normal. The chassis side of suppressors shall be connected as directly as possible to rectifier output ground. Operating characteristics of suppression components provided for power supply rectifier output lines shall be as follows:

(a) <u>MCOV</u>. MCOV shall be above maximum rectifier output voltage.

(b) <u>Leakage current</u>. The DC leakage current shall be less than 1mA for voltages at or below 1.414 x MCOV VDC.

(c) <u>Turn on voltage</u>. Turn on voltage shall be as near MCOV as possible and shall not exceed 125 percent of MCOV.

(d) <u>Discharge (clamp) voltage</u>. Clamp voltage shall be the lowest possible value that can be obtained using suppressors not exceeding 160 percent of MCOV.

(e) <u>Overshoot voltage</u>. Overshoot voltage shall be sufficiently low to preclude electronic equipment damage or operational upset. Time duration of overshoot voltage shall be limited to the shortest possible time not exceeding 2 nanoseconds.

(f) <u>Self-restoring capability</u>. Transient suppressors installed in power supplies shall automatically restore to an off state when line transient voltage falls below rated turnon voltage for the suppressor.

(g) <u>Operating lifetime</u>. The transient suppressors shall safely dissipate 1000 surges with an amplitude of 200 amps and a waveform of 1.2-by-50 μs. One point two (1.2) μs defines the time from the start of the transient to peak amplitude, and 50 μs is the time from the start of the transient until the transient decays to 50 percent of peak value. Methods of testing shall be in accordance with the guidance in IEEE C62.45.

3.6 Conducted Signal, Data and Control Line Transients

Transient protection shall be provided for all signal, data and control lines; both at facility entrances and at entrances to all electronic equipment used in direct support of the NAS. This protection shall limit transients at the equipment entrance to below the equipment susceptibility level. Surge protective devices shall be placed on both ends of signal, data and control lines longer than 10 feet connecting pieces of equipment or facilities, to protect against surges coupled into the wiring or caused by ground reference differences. This includes all signal, data, control, and status lines both internal and external. This also includes interfacility power lines installed above and below grade between facility structures and to externally mounted electronic equipment and particularly vertically routed cables such as those between an ATCT cab and base building or radar tower and base building. These suppressors may be either internal or external to the equipment being protected. All unused conductors of a cable shall be grounded at each end either directly or through an SPD. This shall be accomplished by methods (such as termination on a terminal strip) that allow for future inspection. Additional protection, design and packaging requirements specifically applicable to audio, radio frequency (RF) and other signals transmitted by axial cables are specified in the following subparagraphs. Transient protection shall be provided for all signal, data and control lines including those provided or installed by a telecommunications service provider.

3.6.1 Transient Levels

Electronic equipment using landlines shall be protected against the transient levels defined in Table II. Transient levels for landlines installed in ferrous conduit are different from those for landlines not continuously enclosed in ferrous conduit. The 8-by-1000 μs waveform in Table II defines a transient with an 8 μs rise time and decay to 50 percent of the peak voltage in 1000 μs. Where leased service cables, serving leased equipment, enter the facility, they shall be protected in such a manner that the transient remnant shall not exceed 2.5 times the peak normal signal voltage.

Table II. Conducted Landline Transient Levels

#of Transients (8 – by - 1000μs) Microsecond Waveform	Peak Amplitude (Voltage and Current)			
	Lines Continuously Enclosed In Ferrous Conduit		Lines Not Continuously Enclosed In Ferrous Conduit	
1,000	50V,	10A	100V,	50A
500	75V,	20A	500V,	100A
50	100V,	25A	750V,	375A
5	100V,	50A	1000V,	1000A

3.6.2 Protection Design

Detailed analyses of suppression component and electronic equipment circuit characteristics are required to select components compatible with the requirements herein and to provide suppression circuits that will function without adversely affecting signals and information transmitted by individual landlines. Design requirements for selection of components are as follows:

(a) Unipolar suppression components shall be selected and installed for signals and voltages that are always positive or always negative relative to reference ground. Bipolar suppression components shall be selected for signals and voltages that are both positive and negative relative to reference ground.

(b) The total series impedance of the suppression circuits at both ends of a landline shall be designed so as not to degrade electronic equipment performance.

(c) The high energy protection components at facility entrances shall be selected to reduce the magnitude of transient levels to equipment, clamping or limiting transient parameters safely below electronic equipment susceptibility levels for individual lines.

(d) The suppression components at the facility and electronic equipment entrances, and any impedance added to insure coordination, shall be selected to function together. The transient protection components at the facility entrance and the transient protection components at the electronic equipment entrance shall be selected to clamp and limit the transient voltage and energy safely below circuit susceptibility levels.

3.6.3 Characteristics

The combined operating characteristics for landline transient suppression at facility and electronic equipment entrances and requirements for individual devices shall be as follows:

(a) <u>MCOV</u>. The MCOV rating of the suppression components shall be 15-25% above nominal line voltage or the next higher voltage device available.

(b) <u>Leakage current</u>. Leakage current to ground shall not exceed 100 microamperes at MCOV.

(c) <u>Turnon voltage</u>. Turnon voltage of the suppression components shall be as close to MCOV as possible using state-of-the-art devices, and shall not exceed 125 percent of MCOV.

(d) <u>Overshoot voltage</u>. Overshoot voltage amplitude and duration limits shall be low enough to preclude electronic equipment damage or operational upset. The requirement shall apply for transients with rise times up to 5,000 V/μs.

(d) <u>Clamp (discharge) voltage</u>. Clamp voltage shall be below the electronic equipment susceptibility levels while dissipating the transients listed in Table II.

(e) <u>Operating life</u>. The transient suppression system shall dissipate the transients defined in Table II. Clamp voltage levels shall not change more than 10 percent over the operating life of the suppression system.

(f) <u>Self-restoring capability</u>. The transient suppression system shall automatically return to the off state when the transient voltage level drops below turnon voltage for the suppressors.

3.6.4 Installation of Facility Level Transient Protection

Facility level transient suppression components for signal, data, and control lines shall be installed either at the point where the landlines enter the facility or at the demarc where the lines transfer to FAA control. When a battery feeds signal, data or control lines the suppression components shall be housed in a metal enclosure due to the current available in case of a device failure. When a separate equipment level protector is installed a ground bus bar, electrically isolated from the enclosure, shall be provided to serve as an earth ground point for the facility level transient suppressors. This ground bus bar shall be directly connected to the EES (EES) with an insulated #4 AWG or larger copper conductor of minimum length with no loops, sharp bends or kinks. The conductor insulation shall be color-coded green with a bright red tracer. A UL listed double bolted terminator shall be used to bond the wire to the ground bus bar. The bonding connection to the EES shall be an exothermic weld. The ground bus bar location shall permit a short, direct connection to transient suppressors. The installation shall provide easy access to component terminals for visual inspection, test and replacement. In cases where separate equipment level protection is not required to limit transient voltages to below the equipment susceptibility level; the facility level protection shall be grounded to the multipoint ground system. The location of transient protection for landlines is specified both at entrances to facilities and at entrances to electronic equipment within facilities. Depending on the type of electronic equipment and planned facility installation, combining the transient suppression specified at facility and electronic equipment entrances may be acceptable. Transient protection designs for landlines that combine the protection specified herein for installation at one location shall:
  (1) provide high energy suppression component(s) or device(s) to remove a major percentage of transient energy from each line,
  (2) provide low energy suppression components to reduce transient energy and voltages to below the electronic equipment susceptibility level for each line,
  (3) be located at the entrance to the facility, and
  (4) have approval by the OPI of this document prior to implementation.

3.6.5 Installation of Suppression Components at Electronic Equipment

Equipment level transient suppression components may be housed along with the facility level transient suppression components described in paragraph 3.6.4, or in a separate enclosure, as close to the equipment as possible or as an integral part of the electronic equipment design. Components used shall be leadless or of minimum length with no loops, sharp bends or kinks. Access shall be provided for visual inspection and replacement of components. Equipment level transient suppression components, not housed along with the facility level transient suppression components, shall be grounded to the multipoint ground system as close as possible to the equipment being protected.

3.6.6 Externally Mounted Electronic Equipment

When landlines are directly connected to externally mounted electronic equipment the landline suppression specified in this document for both facility and electronic equipment entrances shall be provided at the equipment entrance. This combined protection shall provide separate high and low energy components with a single grounding path. The grounding conductor shall be bonded to the equipment chassis and shall be of minimum length and routed to avoid sharp bends, kinks or loops. Access shall be provided for visual inspection of these suppressors and for their replacement.

3.6.7 Axial Cables

Transient protection for electronic equipment using coaxial, tri-axial, and twin-axial cables shall be provided both at facility entrances and at the electronic equipment. Transient suppression shall be provided equally for each conductor and shield that is not grounded directly to the electronic equipment case. The protection provided for electronic equipment using axial cables shall comply with the requirements given above as well as the following additional requirements.

3.6.7.1 Protection Design

Special attention shall be given to the design of transient protection for axial-type cables. Design may be particularly critical at RF frequencies. The following design requirements apply.

(a) Suppression circuits shall be designed using state-of-the-art components that have minimum effect upon the signals being transmitted.

(b) Packaging of suppression components and circuits shall be designed to minimize the effect on transmitted signals. Feed through components, leadless components, or short direct lead connections without bends will improve performance of the suppression circuit and reduce signal degradation.

(c) Analyses and tests shall be performed to assure that suppression components do not degrade signals to an unacceptable degree or cause marginal operation of electronic equipment. Particular attention shall be given to the impedance, insertion loss, phase distortion, and voltage standing wave ratio for RF signals.

(d) When transient protection as specified herein cannot be provided without unacceptable degradation of performance, alternatives shall be submitted in writing and implemented with approval of the OPI for this document.

3.6.7.2 Metal Bulkhead Connector Plates

A metal bulkhead connector plate shall be provided where axial-type cables, waveguides, conduits etc. not covered in paragraph 3.2.3 first enter a facility. The connector plate shall be a minimum of 1/4 inch thick, and shall be constructed of tin-plated copper or other material compatible with the connectors. Care shall be exercised to insure that a corrosion situation is not created due to dissimilar metals or a corrosive environment. The plate or plates shall have the required number and types of feed through connectors to terminate all axial lines and provide adequate surface area for bonding waveguides, cable shields, conduits etc. The connectors shall provide a path to ground for cable shields, except when the shield must be isolated for proper equipment operation. If external and internal cables are of different sizes, the changeover in cable size may be accomplished by the feed through connectors at the plate. Waveguides shall be bonded to the bulkhead plates with a minimum #4 AWG conductor; conduits etc shall be bonded to the bulkhead plates with a minimum #4 AWG conductor. The #4 AWG bonding cable for a waveguide can be connected to the waveguide flange with an appropriately sized ring terminal. Conduits shall be bonded with a U-Bolt type bonding connector. Coaxial cable shields shall be bonded with bonding kits sized for the specific cable type. These bonding jumpers shall be connected to the plate with either an exothermic weld or a double bolted lug. The bulkhead plate shall be bonded to the EES with a minimum #4/0 AWG copper cable color-coded green with a red tracer. Additionally, when building steel is properly bonded to the EES, the bulkhead connector plate shall be connected to building steel. Exothermic welds shall be used for these connections.

3.6.7.3 Installation of Suppression Components at Facility Entrances

Transient suppression components for axial-type cables shall be packaged in a sealed metal enclosure with appropriate connectors at each end to permit in-line installation at the bulkhead connector plate required in paragraph 3.6.7.2.

3.6.7.4 Installation of Suppression Components at Electronic Equipment Entrances

Electronic equipment, without adequate internal transient suppression, shall be fitted with suppression components packaged with compatible connectors on each end permitting in-line connection directly to the equipment. When isolation of the cable shield and connector body is required for electronic equipment operation, suppression component grounds shall be isolated from the component housing and connected directly to the electronic equipment case. All electronic equipment case entrances (penetrations) shall be located in a common area, close to the case ground connection point to minimize circulating ground currents on the case.

3.6.8 Fiber Optic Cable

This standard does not preclude the use of fiber optic cable as a substitute for metallic signal, control, and data cables. Where fiber optic cable uses conductive armor, the armor shall be bonded to the EES at the facility entrance. If the cable is internal to the facility, conductive armor shall be bonded to the multipoint ground system at the equipment entrance. The bonding conductor shall be a #2 AWG insulated green with an orange stripe stranded copper conductor. The use of fiber optic cable without a conductive shield or armor is permitted. The transmitter and receiver modules shall be contained in ferrous enclosures, with all penetrations appropriately gasketed or constructed to prevent RF coupling across the shield. Surge protective devices for the metallic signal circuits and power shall be installed as equipment level protection at the fiber optic receiver or transmitter equipment entrance and bonded to the chassis

3.7 Lightning Protection System Requirements

3.7.1 General

The intended purpose of the lightning protection system is to provide preferred paths for lightning discharges to enter or leave the earth without causing facility damage or injury to personnel or equipment. The essential components of a lightning protection system are air terminals, roof and down conductors connecting to the EES, and the EES. These components act together as a system to dissipate lightning energy. The lightning protection system shall meet or exceed the requirements of all FAA standards and orders, Standard for the Installation of Lightning Protection Systems, National Fire Protection Association (NFPA 780), Installation Requirements for Lightning Protection Systems, Underwriters Laboratories (UL 96A) and as specified herein. Ensure that no part of the structure extends outside the zone of protection established by the air terminals or catenary wires. Determine the zone of protection using the criteria in NFPA 780. The provision of a UL Master label is not sufficient to indicate compliance with this FAA standard requirement.

3.7.2 Materials

All equipment shall be UL listed for lightning protection purposes and marked in accordance with UL procedures. All equipment shall be new and of a design and construction to suit the application in accordance with UL 96A requirements, except that aluminum shall only be used on aluminum roofs, aluminum siding or other aluminum surfaces. Bronze and stainless steel may be used for some components. Aluminum materials shall not be used on surfaces coated with alkaline-base paint, or embedded in masonry or cement, on copper roofing, in contact with copper materials, or underground. Bimetallic rated connectors shall be used for interconnecting copper and aluminum conductors. Dissimilar materials shall conform to the bonding requirements of paragraph 3.14.2.3.

3.7.3 Main Conductors

Roof and down conductors shall meet the requirements given in NFPA 780. Roof and down conductors shall be routed outside of any structure and shall not penetrate or invade that structure except as indicated in paragraph 3.7.10.

3.7.4 Hardware

Hardware shall meet the following requirements:

3.7.4.1 Fasteners

Roof and down conductors shall be fastened at intervals not exceeding 3 ft. (0.9 m). Fasteners shall be of the same material as the conductor base material or bracket being fastened, or other equally corrosion resistant material. Galvanized or plated materials shall not be used. Where fasteners are used for bonding the surface shall be prepared in accordance with paragraph 3.14.9.

3.7.4.2 Fittings

Bonding devices, cable splicers, and miscellaneous connectors shall be suitable for use with the installed conductor and shall be copper, bronze or aluminum with bolt pressure connections to the cable. Cast or stamped crimp type fittings shall not be used. Aluminum fittings shall only be used with aluminum conductors. Copper and bronze fittings shall only be used with copper conductors. Interconnection between copper and aluminum portions of the lightning protection system shall be accomplished with bimetallic connectors.

3.7.5 Guards

Guards shall be provided for down conductors located in or next to driveways, walkways or other areas where they may be displaced or damaged. Guards shall extend at least 6 ft. (1.8 m) above and 1 ft. (0.3 m) below grade level. Guards shall be schedule 40 polyvinyl chloride (PVC) pipe where feasible. Metal guards may be used but shall be bonded to the down conductor at both ends of the guard. Bonding jumpers shall be of the same size as the down conductor. PVC guards do not require bonding. Crimp type fittings shall not be used.

3.7.6 Bonds

Certain metallic bodies located outside or inside a structure contribute to lightning hazards because they are grounded or assist in providing a path to ground for lightning currents. Such metallic bodies shall be bonded to the lightning protection system wherever it is likely for a side flash to occur between the lightning protection system conductors and a grounded metal body. As a minimum, this shall be done in accordance with the distance guidance provided in NFPA 780. Bonding should also be applied to other metal bodies, permanently affixed to the structure, because of their size or relative position to the lightning protection system conductor.

3.7.6.1 Metallic Bodies Subject to Direct Lightning Discharge

Metallic bodies, on roofs, subject to direct lightning discharge are generally any large metallic body whose size causes it to protrude beyond the zone of protection (see paragraph 3.7.9) of the installed air terminals. This includes, but is not limited to, exhaust pipes, exhaust fans, metal cooling towers, HVAC units, ladders, railings, antennas, and large louvered structures, etc. When these metallic bodies have a metal thickness of 3/16 in. or greater, they shall be bonded to the nearest main lightning protection system conductor with fittings and conductors that are UL listed for lightning protection. These fittings shall provide bonding surfaces of not less than 3 square inches. Provisions shall be made to prevent corrosive effects introduced by galvanic action of dissimilar metals at bonding points. If the metal parts of these units are less than 3/16 in. thick, additional approved air terminals, conductors and fittings shall be installed, providing an additional path to ground from the air terminals.

3.7.6.2 Metallic Bodies Subject to Induced Charges

Metallic bodies, on or below roof level, that are subject to induced charges from lightning shall be bonded to the lightning protection system in accordance with the distance guidance provided in NFPA 780. This includes, but is not limited to the following metallic items, roof drains, vents, coping, flashing, gutters, downspouts, doors, door and window frames, balcony railing, conduits, pipes, etc. Fittings and conductors used shall be UL listed for lightning protection.

3.7.6.3 Exhaust Stack Grounding.

Exhaust stacks used for either fossil fuel heating or for engine generators create a plume of highly ionized, heated air. This plume attracts cloud to ground lightning. Bond all exhaust stacks to the nearest point in the lightning protection system with a conductor of equal size as the main conductor to prevent flashover. The bond to the exhaust stacks shall be made with an exothermic weld or a mechanical connector UL listed for lightning protection use. Where exhaust stacks are not in close proximity (6 feet) to a main conductor, they shall be bonded directly to a ground rod in the EES.

3.7.6.4   Above Ground Fuel and Oil Storage Tanks.
Provisions shall be made to prevent direct lightning discharge to above ground storage tanks. A separately mounted protection system should be provided where part of the above ground storage tank extends outside the zone of protection provided by the facility lightning protection system. When this is the case, a mast with an air terminal and/or elevated ground wires shall be used. In all cases, mechanical connectors shall be used to bond fuel and vent piping with a 4/0AWG copper conductor or if a 4/0 is not feasible then the largest feasible conductor. This conductor shall be exothermically welded to the EES in accordance with paragraph 3.14.2.1.

3.7.7   Conductor Routing
Roof and down conductors shall maintain a horizontal or downward course. No bend or connection in a roof or down conductor shall form an included angle of less than 90 degrees, nor shall it have a bend radius of less than 8 in. (203 mm). Conductors shall be routed external to buildings and 6 ft. (1.8 m) or more from power (including obstruction lighting power cables) or signal conductors. T-connectors shall not be used to interconnect main conductors. All main conductor transitions shall be swept so as to maintain the 8in or greater bend radius.

3.7.7.1   Down Conductors on Fiberglass Mounting Poles.
Where a fiberglass pole is used to mount an air terminal, the air terminal shall extend at least two feet above the top of the pole and shall be securely fastened to the pole in accordance with the requirements of NFPA 780. The down conductor from the air terminal shall be run on the exterior of the fiberglass pole and shall be fastened to the pole at intervals not exceeding three feet (0.9 m), the down conductor shall be connected to the EES in accordance with paragraph 3.7.8.

3.7.8   Down Conductor Terminations
Down conductors (see paragraph 3.7.3) used to ground air terminals and roof conductors, shall be exothermically welded to a 4/0 AWG copper conductor prior to entering the ground. The 4/0 copper conductor shall enter the ground and be welded to a ground rod that is exothermically welded to the EES. The ground rod shall be located 1 ft. (0.3 m) to 2 ft. (0.6 m) vertically below ground level and from 2 ft. (0.6 m) to 6 ft. (1.8 m) outside the foundation or exterior footing of the building. On buildings with overhangs, ground rods may be located further out.

3.7.9   Buildings
Lightning protection shall be provided for all buildings, or parts thereof, not within a zone of protection provided by another building or higher part of a building, or by an antenna or tower. Zones of protection for all structures shall be as defined in NFPA 780.

3.7.9.1 Air Terminals

Air terminals shall be solid copper, bronze, or aluminum, or, in areas of high corrosion, stainless steel. Copper air terminals may be nickel-plated. Air terminals shall be a minimum of 12 in. (305 mm) in height, at least 1/2 in. (12.7 mm) in diameter for copper and 5/8 in. (15.9 mm) in diameter for aluminum, and shall have a rounded or "blunt" point. Air terminals shall be located in accordance with the requirements of NFPA 780 and UL 96A. Air terminals shall extend at least 10 in. above the object or area it is to protect. Air terminals shall be placed on the ridges of pitched roofs and around the perimeter of flat or gently sloping roofs at intervals not exceeding 20 ft. (6 m) except that air terminals 24 in. (600 mm) or higher may be placed at intervals not exceeding 25 ft. (7.6 m). Air terminals shall be bonded to the nearest roof or down conductor, and connected to the EES in accordance with paragraph 3.7.8. At ATCTs over 100 feet in height, a halo ring shall be provided around the handrail and horizontally mounted air terminals shall be installed at each corner. The air terminals shall be bonded to the halo ring.

> SAFETY NOTE:
> The tip of the air terminals shall not be less than 5 ft. above adjacent walking or working surfaces to avoid the risk of personnel injury.

3.7.9.2 Number of Down Conductors

Buildings less than fifty feet in height, with perimeters of 250 ft. (76 m) or less shall have at least two down conductors. Tall structures, other than antenna towers, above fifty feet in height shall have a minimum of four down conductors. Buildings with perimeters in excess of 250 ft. (76 m) shall have one down conductor for each 100 ft. (30.5 m) of perimeter distance or part thereof. Down conductors shall be as widely separated as possible, i.e. at diagonally opposite corners on square or rectangular buildings The down conductors shall be equally spaced and be continuous from the top of the structure to the EES without any sharp bends, splices or kinks. Building steel and metal supporting structures, or conduits shall not be used in place of down conductors.

3.7.9.3 Metal Parts of Buildings

Metal roofing, structural steel, siding, eave troughs, down spouts, ladders, duct and similar metal parts shall not be used as substitutes for roof or down conductors. A lightning conductor system shall be applied to the metal roof and to the metal siding of a metal clad building in the same manner as on a building without metal covering. Building metal parts shall be bonded in accordance with paragraph 3.7.6.

3.7.9.4 Roof Mounted Antennas

If metallic, the mast of a roof-mounted antenna shall be bonded to the nearest roof or down conductor using fittings and conductors that are UL listed for lightning protection. The bonding jumper shall be of the same size and material as the roof or down conductor to which it is connected. If a roof or down conductor is not available then the antenna mast shall be bonded directly to the EES using fittings and conductors that are UL listed for lightning protection.

3.7.9.5 Radomes Not Mounted On Towers

A 150 ft radius Zone of Protection shall be established to include all Radome surfaces. This shall be in accordance with NFPA-780.

3.7.10 Antenna Towers
Antenna towers shall be provided with lightning protection in accordance with the following:

3.7.10.1 Number of Down Conductors
Pole type towers shall have one down conductor. Towers that consist of multiple, parallel segments or legs that sit on a single pad or footing not over nine square feet in area are considered pole type towers. All other towers shall have at least two down conductors. Large towers, such as radar towers, shall have one down conductor per leg. Down conductors on all towers shall be bonded to each tower section. Down conductors shall be routed down the inside of the legs wherever practical and secured at intervals not exceeding 3 ft. (0.9 m) in accordance with paragraph 3.7.4.1. Down conductors shall terminate at the EES in accordance with paragraph 3.7.8.

3.7.10.2 Towers without Radomes
Pole type towers shall be protected by at least one air terminal to provide a zone of protection for all antennas located on the tower in accordance with NFPA 780. Protection may be provided for large radar antennas by extending structural members above the antenna and mounting the air terminal on top as shown in Figure 1. Structural members shall be braced as required and shall not be used as part of the air terminal or down conductor. The air terminal shall be supported on the structural member and shall have an UL listed fitting on its base. The down conductor from the air terminal shall be connected to a perimeter cable that forms a loop around the perimeter of the tower platform. Down conductors shall be run from the perimeter cable to the EES. Except where only one down conductor is required, each air terminal shall be provided with at least two paths to ground. All conductors shall be in accordance with NFPA 780 requirements for main conductors. All tower legs shall be exothermically bonded to EES with a 4/0 copper cable.

3.7.10.3 Radomes
Radomes shall be located within a zone of protection established according to the 150 ft radius "rolling sphere model" as further described in NFPA-780 paragraph 3.7.3 or FAA Order 6950.19a. This protection can be either from air terminals mounted on the radome or air terminals or catenary wires mounted independently of the radome. When air terminals are mounted on the radome they must have two paths to the EES. An equalization ring shall be established at the radar antenna deck level.

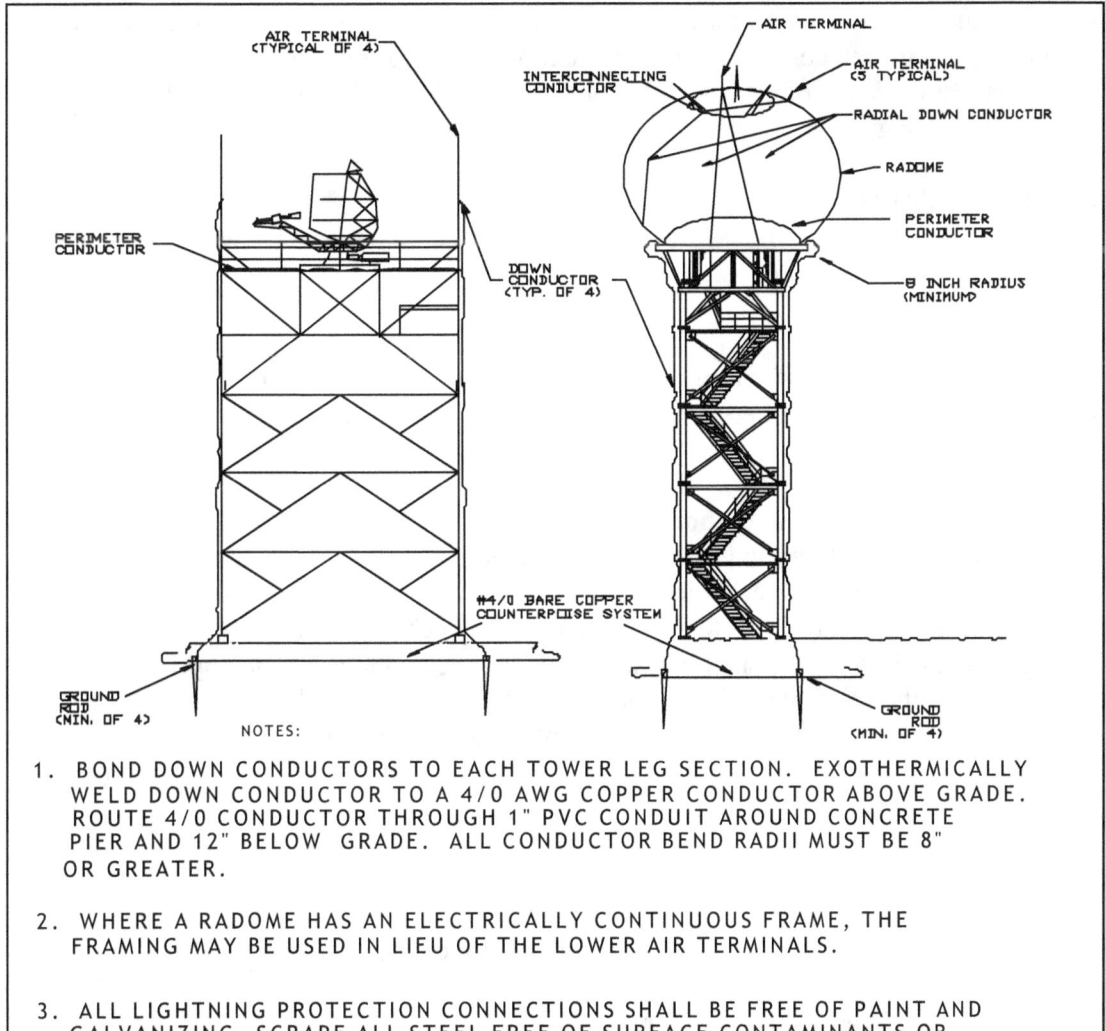

Figure 1. Lightning Protection for Radomes and Radar Antenna Platforms

3.7.10.4 Towers with Radomes

Towers with spherical radomes shall be protected with a 2 ft. (0.62 m) air terminal at the peak and four or more air terminals equally spaced around the circumference of the radome. The spacing and quantity of the circumferential air terminals may be adjusted if the antenna pattern is affected, but their sizing, position and height shall establish a protection zone as specified in 3.7.10.3. The circumferential air terminals shall be interconnected with a main sized conductor. This conductor, labeled "Interconnecting Conductor" as indicated in Figure 1, shall be connected to the air terminal on the peak. The "Interconnecting Conductor" shall also be connected to the perimeter cable that forms a loop around the base of the radome. These interconnecting conductors shall be run from the air terminal at the peak of the radome, in a path following the contour of the radome, to the perimeter cable as shown in Figure 1. Deviations from the shortest possible path may be used where near field radar analyses determine that interference from the conductors will degrade the performance of the radar. Any bends in the interconnecting conductors shall maintain the largest possible radii and in no case be less than 12 inches. One down conductor per leg shall connect the perimeter cable at the base of the radome to the EES. These down conductors shall be bonded to each leg section and exothermically welded to the lowest section of each leg.

3.7.10.5 Antenna Protection
Air terminals shall be placed to protect structural towers and buildings, and antennas mounted to towers and on buildings.

3.7.10.6 Tower Guying
All metallic guy wire systems without insulators shall be connected to the EES. Cathodic protection for metallic anchors, either active or passive is advisable at facilities particularly those with a large EES.

3.7.10.6.1 Low Conductivity Anchors
On guy wires terminating in low conductivity anchors (such as concrete), a jumper of the same material as the guy wire shall be mechanically bonded to each guy wire above its lowest turnbuckle. Where multiple guy wires terminate on a single anchor, the jumper may daisy chain through the guy wires. The jumper shall be exothermically welded to a 3/4 inch by 10foot ground rod that is exothermically welded to the EES. Mechanically bonded jumpers of the same material as the guy wire shall be placed across any intermediate turnbuckles in a guy wire. All jumper connections to the guy wires shall be made with appropriate compatible connectors that do not create a corrosion cell.

3.7.10.6.2 Metallic Anchors
A jumper, of the same material as the guy wire, shall be bonded across each turnbuckle in the guy wire with appropriate compatible connectors that do not create a corrosion cell.

3.7.10.7 Waveguide, Axial Line, and Conduit Protection
Waveguide, axial line, and conduit located on the tower and feeding into the facility shall be separately bonded to a ground plate mounted on the tower or directly to the EES. This bond shall be at a point no greater than 2 feet above the transition bend (90 degree bend) near the tower's base. The ground plate shall be bonded to the EES with a #4/0 AWG copper cable in accordance with the requirement in paragraph 3.6.7.2. A separate bond shall be made from the point of origin within the tower structure of each waveguide, axial line, or conduit to the metallic tower structure.

3.7.10.8 Staircase/Ladder Protection
The metallic access, i.e., staircase, ladder, etc., shall be exothermically bonded, near its base, to the EES with a #4/0 AWG copper conductor installed in a location that prevents accidental trips or strikes that could result in personnel injury. Where staircase sections are not welded together, bonding jumpers shall be installed between sections.

3.7.11   Fences

Grounding and bonding of fences shall be done to all areas where personnel may come in contact with the metallic fence structure or fabric or where a significant step or touch potential may be present.

3.7.11.1 Fences and Gates Requiring a Counterpoise

Any fence that is within a horizontal distance equal to one and one-half times the height of the tallest structure shall be bonded and grounded as shown in Figure 2. Fences made of conducting material, i.e. chain link fabric, metal crossbar, stranded wire, shall be constructed using metal posts which extend a minimum of 2 ft. (0.6 m) below grade. Gates shall have a 1 in. by 1/8 in. flexible tinned copper bond strap or an insulated #4/0 AWG flexible (welding) copper cable that is bonded to the adjacent post (exothermic welding is recommended). The posts at each side of the gate shall also be exothermically welded, at a height no greater than 1 foot above grade, to their respective ground rods with a #4/0 AWG bare copper cable. The bonding strap to the post shall be installed so as not to limit full motion of the gate (whether swing or slide type). An exothermically welded 4/0 AWG bare copper conductor shall also connect the ground rods to the EES. Metallic fence fabric with non-conductive coatings shall not be used. A horizontal bare #6 AWG stranded copper conductor, shall be woven through the fencing fabric and shall be mechanically bonded to the fence posts at intervals not greater than 40 feet. The fence posts with these bonds and fence posts adjacent to gates shall be bonded to the EES with a #4/0 AWG bare copper cable exothermically welded at each end. Additionally, a horizontal bare #6 AWG stranded copper conductor shall be woven continuously through the gate fabric and mechanically bonded to the gate rails. The method of bonding a fence is illustrated in Figure 2. This method requires a counterpoise around the facility and an additional outer counterpoise around and within 3 feet of the fence. The two counterpoise systems shall be connected together in as many places as possible (4 minimum for a small facility <75 ft. square, 8 minimum for a large facility >75 ft. square) to further equalize the step potentials within the facility.

3.7.11.2 Fences and Gates Not Requiring a Counterpoise

When a fence constructed of conducting materials is not located within a horizontal distance equal to one and one-half times the height of the tallest structure it shall be constructed using metal posts that extend a minimum of 2 ft. (0.6 m) below grade. Gates shall have a 1 in. by 1/8 in. flexible tinned copper bond strap or an insulated #4/0 AWG flexible (welding) copper cable that is bonded to the adjacent post (exothermic welding is recommended). The posts at each side of the gate shall also be exothermically welded, at a height no greater than 1 foot above grade, to their respective ground rods with a #4/0 AWG bare copper cable. The bonding strap to the post shall be installed so as not to limit full motion of the gate (whether swing or slide type). A #4/0 AWG cable shall interconnect the ground rods at either side of the gate.

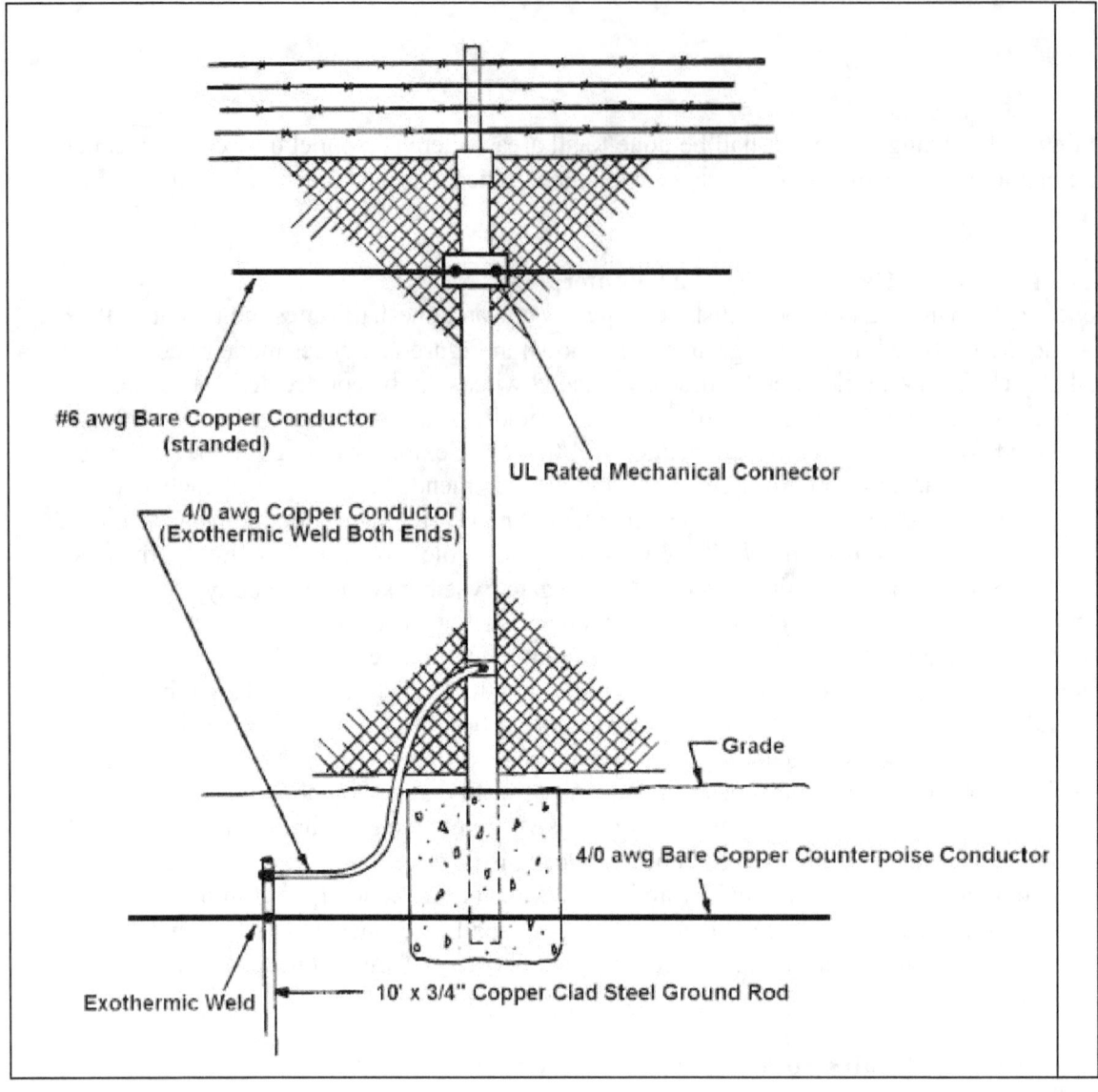

Figure 2. Grounding a Fence

3.7.11.3 Overhead Power Line Considerations

When a fence constructed of conducting material is crossed by overhead power lines, and is not within a horizontal distance equal to one and one-half times the height of the tallest structure, the fence shall be bonded on each side of the crossing to a ground rod with a bare #4 AWG solid copper protective conductor as shown in Figure 2. These protective-grounding connections shall be a minimum of 20 ft. from the overhead wire crossing and on each side of the crossing. For a chain link fence, the fabric shall be bonded to these conductors at the top, middle and bottom and at each strand of security wire placed above the fencing fabric. Where cross bars or stranded wire is used, each horizontal strand or cross bar shall be bonded to the conductors. Ends of the protective conductors shall be connected, using exothermic welds to their respective ground rods and the other ends of the conductors shall be connected to the top of the fence.

3.7.12 Airport Traffic Control Towers (ATCT).

ATCTs, as shown in Figure 3 having electronic areas in the cab, junction and sub-junction levels at the top of the shaft and also in the associated base building present a unique set of challenges for implementing lightning and transient protection. The numerous conductors running between electronic equipment located in the base building and beneath the tower cab are subject to large electromagnetic fields during a lightning strike. For this reason, special techniques must be applied to provide an environment that minimizes the damaging effects of lightning. These techniques are mandatory for ATCT facilities over 100' in height with base buildings, and in isokeraunic areas of 30 thunderstorm days annually or greater.

3.7.12.1 General.

The lightning protection, electrical, electromechanical, electronic systems, and building steel of structures must be bonded together for safety. The National Electrical Code (NEC) NFPA-70 as well as this and other FAA Standards and Orders mandate this bonding. It is not possible for equipment near the top of the tower and at the base to have the same electrical potential during a lightning strike. It is therefore necessary to reference all systems at the top of the tower to each other and treat this area as a separate facility.

3.7.12.2 Main Ground Plate and Power Distribution.

In order to assure good high frequency grounding during normal operation a low impedance connection must be provided to the EES. A main ground plate shall be established on the lowest level with electrical, electromechanical, or electronic equipment serving the ATCT cab (See Figure 3). All grounding systems present at or above this level within the ATCT shall be connected to this main ground plate. A 1-foot wide #26 gauge or thicker copper strap shall connect this main ground plate to a plate at the base of the ATCT. This base plate shall be grounded via 2 ea. 500 kcmil cables exothermically welded to the base plate and to the EES.

This strap will provide 2 ft^2 of surface per lineal foot of conductor and shall be routed continuously from the main ground plate to the base plate without sharp bends, loops, kinks, or splices. A combination of smaller conductors providing the same surface area per lineal foot may be substituted. This conductor shall be mechanically bonded to the main ground plate and the base plate. The strap should be sandwiched between the plate at each end and a 1'x1"x1/8" copper backing to insure good electrical contact and mechanical strength. The connection from the base plate to the EES shall be accomplished in an access well to facilitate periodic inspection. All power distribution for the areas at the top of the ATCT shall be via separately derived source(s). These separately derived source(s) shall be grounded in accordance with the requirements of NEC article 250. The Grounding Electrode Conductor (GEC) specified in NEC article 250 shall be connected to the grounded and grounding conductors at the first system disconnecting means or overcurrent device. This point of connection is mandated to facilitate the effective installation of an SPD. An SPD rated at 80kA (8x20 microsecond current waveform) surge capability or greater, suitable for location category C3 per IEEE C62.41-1991, and providing protection L-L and L-N shall be installed on the load side of the first disconnecting means or overcurrent device of each separately derived system. The ground bus at the first disconnecting means or overcurrent device shall be bonded to the junction level main ground plate established in accordance with the requirements of this paragraph. This connection shall not be in lieu of the grounding electrode conductor requirements of NEC article 250.

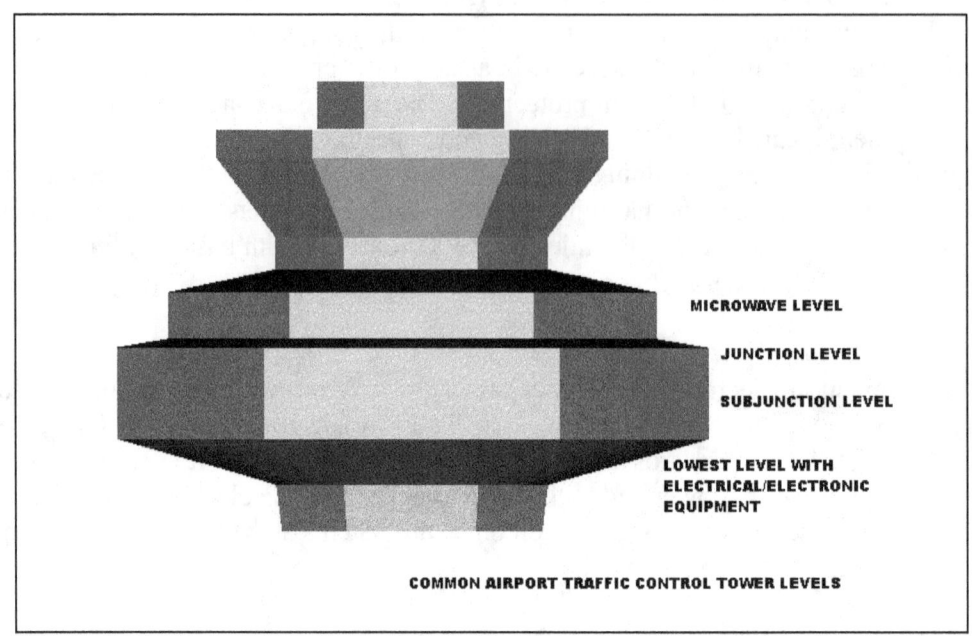

Figure 3. Airport Traffic Control Tower Levels

3.7.12.3 Roof, Structural Steel, Reinforcing, and Other Metal Element Bonding.
Metal elements composing the ATCT roof and its supporting structure, reinforcing bar (rebar) in both horizontal and vertical elements, building steel, and metal sheathing shall be bonded together so as to provide a "Faraday Cage". Particular care shall be taken to insure that all penetrations of the Faraday cage are bonded to the cage at their point of entry. All rebar within the tower shaft shall be tied together and where precast reinforced concrete panels are used, the rebar shall be tied between panels at least once per 4 feet. Rebar in the floors, overheads, corrugated decking, and footers shall likewise be tied to the rebar in the vertical elements. These ties serve as a means of bonding of the rebar. This bonding is necessary to establish both the Faraday Cage and to provide a secondary-grounding path for high frequency equipment.

3.7.12.4 Signal, Communications, Axial Cables and Control Line Protection.
For purposes of lightning and transient protection cables running up the tower shaft in open raceways are considered totally exposed to lightning related phenomenon. For this reason transient protection must be applied at each end of these cables. This protection shall be installed where the cables enter the equipment room near the top of the ATCT and where they enter the associated base building. Cables between the equipment rooms and tower cab area shall be protected in accordance with paragraph 3.6. Both facility and equipment levels of protection shall be provided for these lines. Enclosing metallic cabling in RGS or the use of all dielectric fiber optic cable can significantly reduce the threat of lightning related damage to ATCT and base building circuits.

3.7.12.5 Signal Grounding.
The signal grounding system for the ATCT cab and associated electronic equipment rooms consist of both single point and multipoint elements. The single point grounding system is most frequently used in conjunction with the audio and associated switching equipment. The multipoint ground system is used for most other electronic equipment. All grounds present within the operational or equipment levels shall be bonded together on the lowest level with electrical, electromechanical, or electronic equipment serving the ATCT cab (See Figure 3).

3.7.12.6 Multipoint Ground.
A multipoint ground system consisting of either a raised access floor with below floor signal reference grid (SRG) or, where a raised access floor is not used, a copper sheet equipotential plane (EPP) shall be installed in:
a) All facility operational equipment areas
b) All other areas containing electronic equipment supporting facility operations.
The above operational and electronic equipment - and all electrical equipment in those areas - shall be bonded to the SRG or EPP installations in the area. In turn, all installed SRG's and EPP's - on the same floor and on different floors - shall be bonded together.
c) Any area containing electrical equipment installed to address power quality (e.g., isolation transformers, power conditioning equipment, etc.) not in the same area as the operational or electronic equipment (on different floors, etc.) shall be bonded to the SRG/EPP system.

The SRG shall consist of 2" wide copper strips arranged in a grid on 2' centers. Connections from the SRG to the access floor pedestals shall be on a six foot spacing. The EPP shall be either in or on the floor. Floor coverings of either tile or carpeting shall be of static dissipative material that is properly installed per manufacturers' specifications and connected to a component of the Multipoint Ground (MPG) system or to the Signal Reference Grid (SRG). The floor covering should have a surface-to-surface resistivity (R_{tt}) of between 25 kΩ (2.5×10^4) per square (minimum) and 100 MΩ (1×10^8) per square (maximum) and be tested at a minimum semiannually in accordance with the test method specified in ANSI/ESD S7.1-1994, "Floor Materials --Resistive Characterization of Materials." Individual areas of the multipoint ground system on a single floor shall be bonded to adjacent areas via at least two separate paths providing a minimum of 2 ft^2 of surface area per lineal foot of conductor per path. The grounding system on each floor with electrical, electromechanical, or electronic equipment shall be bonded to adjacent floors via at least two separate paths providing a minimum of 2 ft^2 of surface area per lineal foot of conductor per path.

3.7.12.7 Single Point Grounding.
Single point ground systems, if required for the electronic equipment to be installed, shall be constructed in accordance with paragraph 3.11.7.1 of this document. All single point ground systems and independent ground systems mandated by equipment manufacturers shall be bonded to the ATCT junction level main ground plate established in accordance with the requirements of this paragraph. All electronic grounding systems at a facility shall be bonded together to prevent the possibility of large voltage differentials between equipment during a lightning strike.

3.8 Earth Electrode System (EES) Requirements

3.8.1 General

An EES shall be installed at each facility to provide a low resistance to earth for lightning discharges, electrical and electronic equipment grounding, power fault currents and surge and transient protection. The EES shall be capable of dissipating within the earth the energy of direct lightning strikes with no ensuing degradation to itself. The system shall dissipate DC, AC and RF currents from equipment and facility grounding conductors.

3.8.2 Site Survey

A site survey shall be conducted for all sites to determine the geological and other physical characteristics. Information to be collected shall include location of rock formations, gravel deposits, soil types etc. A soil resistivity test shall be performed at distances of 10, 20, 30 and 40 foot (3,6,9 and 12m) spacing in four directions from the proposed facility. All survey data, including soil restivity measurements, shall be noted on a scaled drawing or sketch of the site and included in the facility reference data file (FRDF). Additional guidance may be found in FAA Orders 6950.19 and 6950.20.

3.8.3 Design

The EES shall normally consist of driven ground rods, buried interconnecting cables and connections to underground metallic pipes, tanks and structural members of buildings that are effectively grounded. The site survey required in Paragraph 3.9.2 shall be used as the basis for the design of the EES. The design goal for the resistance to earth of the EES shall be as low as practicable and not over 10 ohms. Where conditions are encountered, such as rock near the surface, shallow soils, permafrost and soils with low moisture or mineral content the ground enhancements listed in paragraphs 3.8.3.1 through 3.8.3.5 may be necessary.

3.8.3.1 Chemical Enhancements.

Chemical enhancements (doping) with materials such as mineral salts, Epsom salts, sulfates, etc. should only be utilized as a last resort. Chemical enhancement is dependent on soil moisture content and requires periodic (usually yearly) re-treatment and continuous monitoring to be effective. The chemicals leach into the surrounding soil and can be deposited into the water table. Typical installation is in bored holes with ground rods and in trenches.

3.8.3.2 Chemical Rods.

Chemical rods also require re-treatment and monitoring to ensure continuous effectiveness. Many of these systems require a drip irrigation system in dry soil conditions. Inspections must be done frequently for excessive corrosion at connection points between the conductor and the chemical rod attachment point. Normal installation is insertion into the soil in accordance with manufacturer's instructions.

3.8.3.3 Engineered Soils.

Engineered soils are soils or clays treated with a variety of materials to enhance their conductive properties. These engineered soils may be a mixture of moisture absorbing materials such as Bentonite, homogenous clays in combination with native soils and/or chemicals. Some engineered soil enhancements utilize ferro-concrete based materials. These materials should be avoided in areas with soil movement. The concrete can break the interconnecting conductor when combined with soil movement. Engineered soils require the presence of moisture (> 14%) in the soil to be effective. The ferro-concrete type enhancement can be very expensive. Normal installation is installation in bored holes with ground rods and in trenches.

3.8.3.4 Ground Dissipation Plates

In shallow soil locations with limited surface space, ground dissipation plates must be installed to supplement the earth electrode system. The plates, installed at the corners of the EES, provide a large surface area of high conductivity at the farthest accessible point from the facility to be protected. This allows the available energy to travel away from the facility. Plates should be constructed of one quarter-inch thick copper and be a minimum of two feet square. These plates should be installed in a vertical plane to take advantage of seasonal moisture and temperature changes in the soil. Install the plates at the same depth or deeper than the interconnecting conductor, but maintain a minimum of one-foot of native soil above the upper edge of the plate. Attachment to the EES should be with a 4/0 AWG 7-strand copper conductor, exothermically welded to the EES and the plate. For maximum performance, the attachment point at the plate should be at the center of the plate, not near the edge or the corners. To further enhance the effectiveness of ground dissipation plates, they may be configured as a Jordan Dissipation Plate, (JDP®) with "pinking shear edges" as in Figure 4. This configuration provides 2/3 more surface area at the edge than a square plate and provides multiple sharp points for increased dissipation capability. In difficult soils/areas a combination of coke breeze trenches and dissipative plates is highly recommended (see Figure 4 and Figure 5).

3.8.3.5 Coke Breeze

Coke breeze is a material that is produced as a by-product of coke production. Coke is used in the smelting of metals. It is mainly fixed base carbon. All the corrosives and volatiles have been cooked off at extremely high temperatures. Coke breeze is environmentally safe, stable, and conductive when completely dry or frozen, non-moisture dependant, compactable and very economical to install. Normal installation is in a one-foot square trench in a counterpoise configuration with a continuous 4/0 AWG 7-strand copper conductor in the center of the material (see Figure 5). Placement of the trench is based on the geometry of the facility and the physical site location. Radial trenches with a center conductor can be utilized to enhance Radio Frequency (RF) ground planes in communication facilities. The top of the coke breeze trench must be covered by a minimum of one foot of native soil. Neither charcoal nor petroleum based coke breeze may be substituted for coke breeze derived from coal in coke ovens. Charcoal and petroleum coke typically contains high levels of sulfur, which in the presence of moisture will accelerate corrosion.

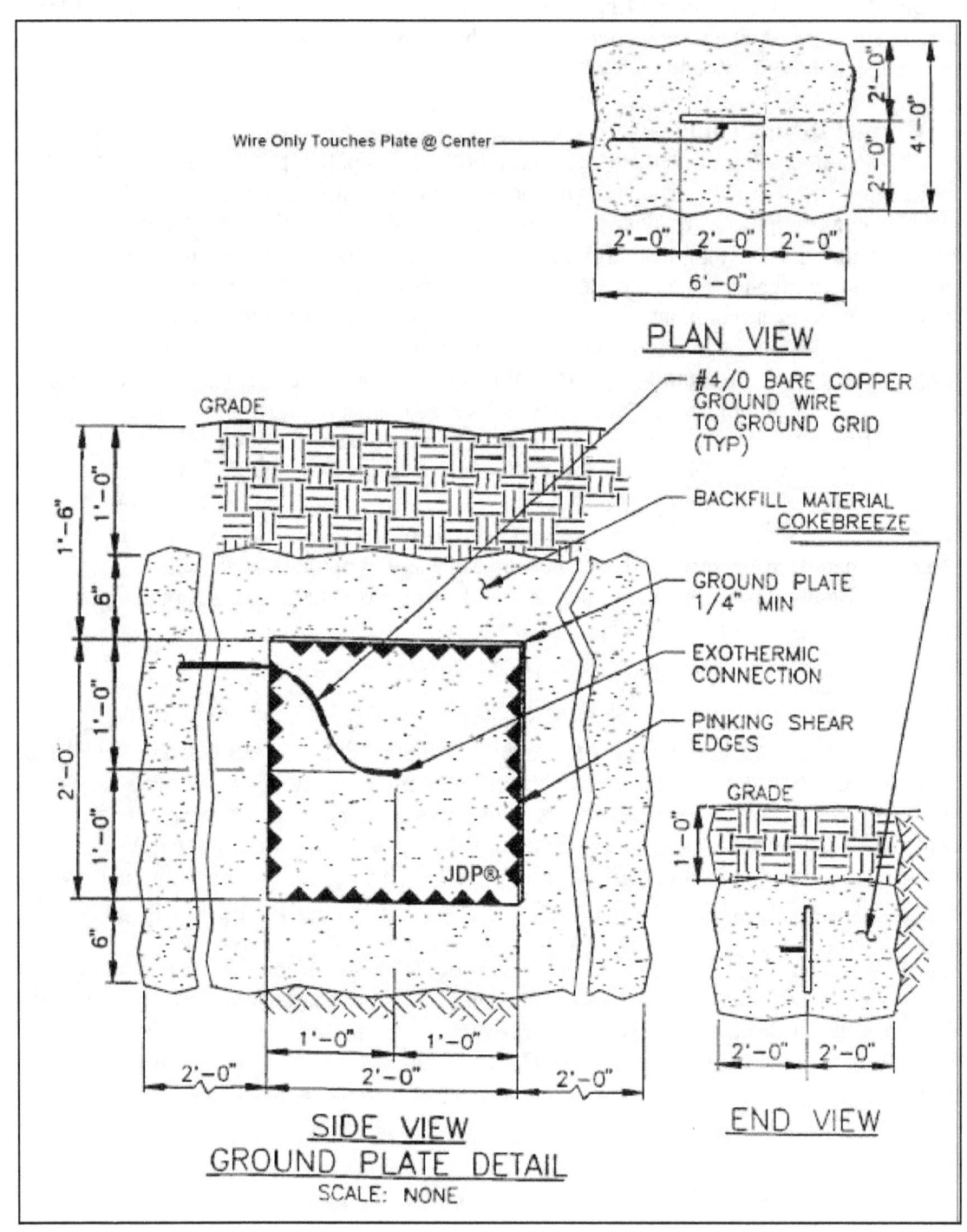

Figure 4. Ground Plate Detail

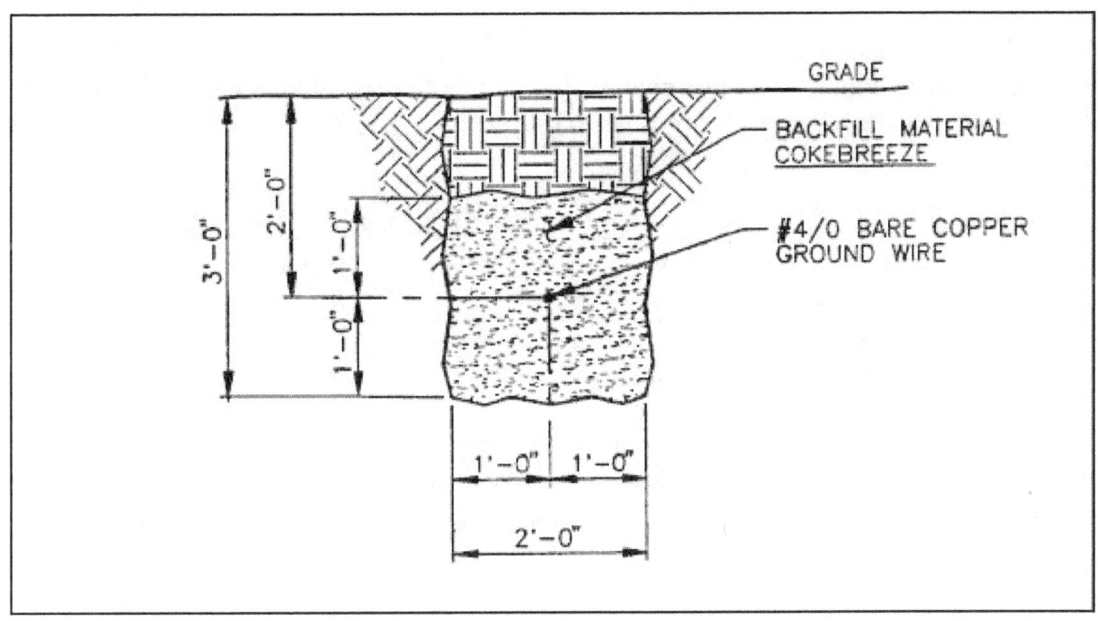

Figure 5. Grounding Trench Detail

3.8.3.6 Installation of Earth Electrode Systems in Corrosive Soils

Careful consideration must be given to the installation of any grounding system in soils with corrosive elements. Two geological areas of known concern are the volcanic soils in Hawaii and Alaska. It is recommended that supplemental cathodic protection be applied to the grounding system at these locations. A buried steel plate (acting as a sacrificial anode) is connected to the EES by a 7-strand 4/0AWG bare copper conductor. The 4/0 conductor should be exothermically welded to the EES and to the sacrificial plate. The conductor should be welded to the center of the plate, not near the edge or near the corners. Minimum sizing for the sacrificial plate is four foot square (4'x4') @ ½ inch thickness. In shallow soils, this would be in addition to the standard copper dissipation plates. For enhanced performance sacrificial plates may be configured as a JDP® (see Figure 4).

3.8.4 Configuration

The EES shall consist of at least four ground rods whose configuration and depth shall be determined by a soil test included in the site survey. At facilities that have two or more structures, i.e. a building and antenna tower, separated by 15-ft. (4.5 m) or less, a single EES surrounding both structures shall be provided. Where structures are separated by more than 15-ft. (4.6 m) but less than 30-ft. (8.2 m), the EESs may share a common side. Where the structures are separated by more than 30-ft. (8.2 m) an EES shall surround each structure and the EESs shall be interconnected by at least two buried cables. Guidelines are provided in FAA Orders 6950.19 and 6950.20.

3.8.5 Ground rods

Ground rods and their installation shall meet the following requirements:

3.8.5.1 Material and Size
Ground rods shall be copper or copper clad steel, a minimum of 10 ft. in length and 3/4 in. in diameter. Rod cladding shall not be less than 1/100 in. thick.

3.8.5.2 Spacing
Ground rods shall be as widely spaced as possible, and in no case spaced less than one rod length. Nominal spacing between rods should be between two and three times rod length.

3.8.5.3 Depth of Rods
Tops of ground rods shall be not less than 1-ft. (0.3 m) below grade level.

3.8.5.4 Location
Ground rods shall be located 2 to 6 ft. (0.6 to 1.8 m) outside the foundation or exterior footing of the structure. On buildings with overhangs, ground rods may be located further out.

3.8.6   Interconnections
Ground rods shall be interconnected by a buried, bare, #4/0 AWG 7-strand copper cable. The cable shall be buried at least 2 ft. (0.6 m) below grade level. Connections to the ground rods shall be exothermically welded. The interconnecting cable shall close on itself forming a complete loop with the ends exothermically welded. The structural steel of buildings shall be connected to the EES at approximately every other column at intervals not over 60 ft. (18.3 m) with a bare, #4/0 AWG 7-strand copper cable. Connections shall be by exothermic welds. The grounding electrode conductor (GEC) for the electric service, sized in accordance with the NEC requirement for grounding electrode conductors, but shall not be smaller than #2 AWG and shall be connected to a ground rod in the EES with an exothermic weld in accordance with paragraph 3.12.2(a). For services greater than 200 amps, the minimum size of the GEC shall be #4/0 AWG copper cable. All underground metallic pipes and tanks (unless cathodically protected), and the telephone ground, if present, shall be connected to the EES by a copper cable no smaller than #2 AWG. Where routed underground, interconnecting cables shall be bare. Exothermic welds shall not be used where hazards exist, i.e. near fuel tanks. In these cases, connections using 14-ton pressure connectors will be allowed. Bonding resistance of all interconnections shall be one (l) milliohm or less for each bond when measured with a 4-terminal milliohm meter.

3.8.7   Access Well
Access wells are permissible at facilities. The well should be located at a ground rod that is in an area with access to the open soil so that checks of the EES can be made once the facility is in use. The access well shall be made from clay pipe, poured concrete, or other approved wall material and shall have a removable cover. The access well shall be constructed to provide a minimum clearance (12 inches radius) from the center of the ground rod to the inside wall of the access well. The opening shall equal or exceed the minimum clearance required for the access well. Connections shall be by exothermic welds.

3.9   Main Ground Plate
A Main Ground Plate shall be established as the main collection plate for the multipoint ground system and the single point ground system for the entire facility. This Main Ground Plate shall

be connected to the EES via 2 ea. 500 kcmil cables. The cables from the main ground plate to the EES shall be exothermically welded at the EES and shall be exothermically welded or connected with UL listed pressure connectors to the plate. The Ground plate location shall be chosen to minimize cable length, but shall not exceed 50 feet. Ground plates shall be copper and at least 12 in. (305 mm) long, 6 in. (152 mm) wide and 1/4 in. (6.4 mm) thick. The Main Ground Plate shall have a clear plastic cover that bears the caption "MAIN GROUND PLATE" in black 3/8-in high (10mm) characters and green slashes around the caption. The Main Ground Cable shall have a solid green color.

3.10 Electronic Multipoint Ground System Requirements

3.10.1 General

All FAA enclosed building facilities used in the operation of the NAS shall have a multipoint ground system. The protection of electronic equipment against potential differences and static charge buildup shall be provided by interconnecting all non-current-carrying metal objects to an electronic multipoint ground system that is effectively connected to the EES. The multipoint ground for electronic equipment systems consists of electronic equipment, racks, frames, cabinets, conduits, raceways, wireways, cable trays enclosing electronic conductors, structural steel members, and conductors used for interconnections. The electronic multipoint ground system shall provide multiple low impedance paths to the EES as well as between various parts of the facility, and the electronic equipment within the facility so that any point within the system is tightly connected (electrically speaking) to the EES. This will minimize the effects of spurious currents that may be present in the ground system due to equipment operation or malfunction, or from lightning discharges. The multipoint ground system shall not be used in lieu of the safety ground required by the NEC. Single point electronic grounds shall not connect to the multipoint ground system, except as specifically permitted by paragraph 3.11. The multipoint ground system is not to be used as a signal return path. A typical ground system is shown in Figure 6. Facility Grounding System.

Exception: For enclosures housing facilities or equipment 100 ft^2 or less in area, a multipoint ground system is not required. Instead, a main ground plate shall be established and connected to the EES with a 4/0AWG conductor. All signal grounding (single point or multipoint) shall terminate on this point.

3.10.2 Facilities Requiring a Signal Reference Structure (SRS)

All new facilities, whose operational and electronic equipment areas exceed a total of 400 ft^2, shall be equipped with a Signal Reference Structure (SRS) installed in accordance with paragraph 3.10.2.1. This SRS shall serve as the multipoint ground system for those areas so equipped. As areas are renovated in facilities meeting the above area criterion they shall be equipped with an SRS. In areas where an SRS is installed in accordance with the requirements of this paragraph, multipoint ground plates are not required.

3.10.2.1 Equipotential Planes in New Facilities

The SRS to be installed in new facilities shall consist of a grid of 2" wide, thin (26 gauge or thicker) copper strips, laid on a 2' grid, below a rigid or bolted stringer, computer access floor. The grid and access floor shall be bonded together at least every 6'. A #4/0 AWG bare copper cable shall be run around the perimeter of all areas, within 6"of the wall, where this SRS is utilized. Bond this perimeter cable to the below floor grid at every intersection. All building structural steel, such as columns, within 6 feet of the grid shall be bonded to the grid with a #4/0 AWG or larger conductor. All concrete encased steel, in new FAA construction, shall be equipped with a grounding terminal. All conduits, wireways, pipes, cable trays, or other metallic elements that penetrate the area shall be bonded to the grid where they enter the area and every 25 feet for their entire length. All conduits, wireways, pipes, cable trays, or other metallic elements within 6 feet of the grid shall be bonded to the grid. The concrete floor beneath the copper grid shall be prepared by coating the floor with white epoxy paint. This coating helps to limit corrosion of the grid and improves below floor visibility. Thus, an equipotential plane is created by the interconnection of the access floor system (with metal backed panels); below floor grid; structural steel elements and electrical supporting structures.

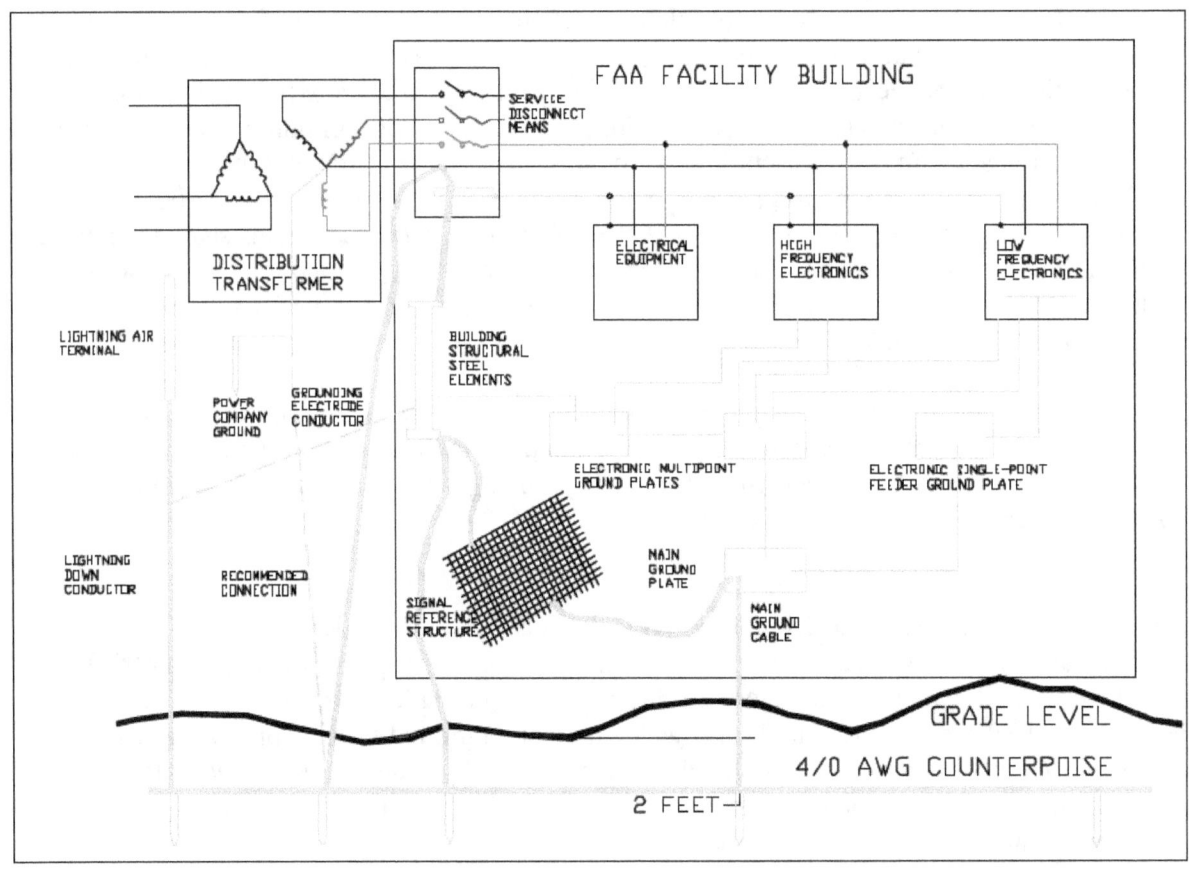

Figure 6. Facility Grounding System

3.10.2.2 SRS Methods in Existing Facilities
In existing FAA facilities it may not be feasible to implement the techniques used to install an equipotential plane as specified the paragraph 3.10.2.1. In these cases the alternative methods described in IEEE 1100-1999 paragraph 8.5.4 and its subparagraphs shall be used.

3.10.2.3 Bonding Electronic Equipment to The SRS
Signal grounding connections for all equipment, especially electronic equipment, to the SRS, may be either to the below floor grid or directly to the access floor system or alternatively to the SRS as constructed in paragraph 3.10.2.1. To prevent the possibility of problems due to resonance of a single bonding strap, two widely spaced straps of unequal length shall be used to connect the equipment to the SRS. Bonding straps shall be at least ¾" wide and at least #26 gauge. Bonding shall be in accordance with paragraph 3.14.3.

3.10.2.4 Facilities without SRS
New facilities which do not meet the area criterion in paragraph 3.10.2 and existing FAA facilities which have not been upgraded shall be equipped with a multipoint ground system consisting of cables and plates in accordance with paragraph 3.10.3.

3.10.3 Ground Plates, Cables and Protection
The electronic multipoint ground system shall not replace the NEC mandated equipment grounding conductor routed with the phase and neutral conductors. At least two connections between the multipoint ground system and the EES shall be provided. One connection shall be provided by a 500 kcmil or equivalent copper cable connected between the main ground plate and the EES. A 500 kcmil or equivalent copper cable, connected between the EES and a supplemental ground plate on the opposite side of the facility shall provide a second connection to the EES. A supplemental ground plate, exothermically welded to building steel and grounded in accordance with paragraph 3.8.6 may be substituted for this second connection to the EES. The cables from the main ground plate and the supplemental ground plate to the EES shall be exothermically welded at the EES and shall be exothermically welded or connected with UL listed pressure connectors to the plates. Connection points shall be chosen to minimize cable length, but shall not exceed 50 feet. In steel structures, additional connections shall be made between each ground plate or bus and the structural steel.

3.10.3.1 Ground Plates and Buses
A ground plate shall be used when a centralized connection point is desired. The location shall be chosen to facilitate the interconnection of all equipment cabinets, racks and cases within a particular area. If more than one ground plate is required, they shall be installed at various locations within the facility. Ground buses shall be used when distributed grounding is desired with a long row of equipment cabinets. Ground plates shall be copper and at least 12 in. (305 mm) long, 6 in. (152 mm) wide and 1/4 in. (6.4 mm) thick. Ground bus width and thickness shall be selected from Table III, Size of Electronic Multipoint Ground Cables, according to the length required. Ground plates and buses shall be identified with a permanently attached plastic or metal label that is predominantly green with distinguishing bright orange slashes. The label shall bear the caption "ELECTRONIC MULTIPOINT GROUND SYSTEM" in black 3/8-in. high (10 mm) characters.

3.10.3.2 Ground Conductors (Plate to Plate and Plate to Bus)

Conductors between plates and buses in the multipoint system shall be sized in accordance with Table III based on the maximum path length to the farthest point in the multipoint ground system from the EES. To determine the distance to the farthest point in the multipoint system, add the length of all cables in the multipoint system to reach the farthest plate in the system via the longest path as shown in Figure 7. Divide the sum obtained by two to obtain the maximum path length. Utilize this path length to determine the conductor size from Table III, but in no case smaller than #4/0 AWG. These conductors shall be color coded green with a bright orange tracer or shall be clearly marked for 4 in. at each end and wherever exposed with a green tape overlaid with a bright orange tracer. Where routed through raceways or wireways, the color-coding shall be visible by opening any cover. Where conductors are routed through cable trays, color-coding 4 in. long shall be provided at intervals not exceeding 3 ft. The use of uninsulated (bare) conductors are not permitted inside structures, except when used for short grounding jumpers, bonding jumpers, and similar items, that are not enclosed in conduit or raceway.

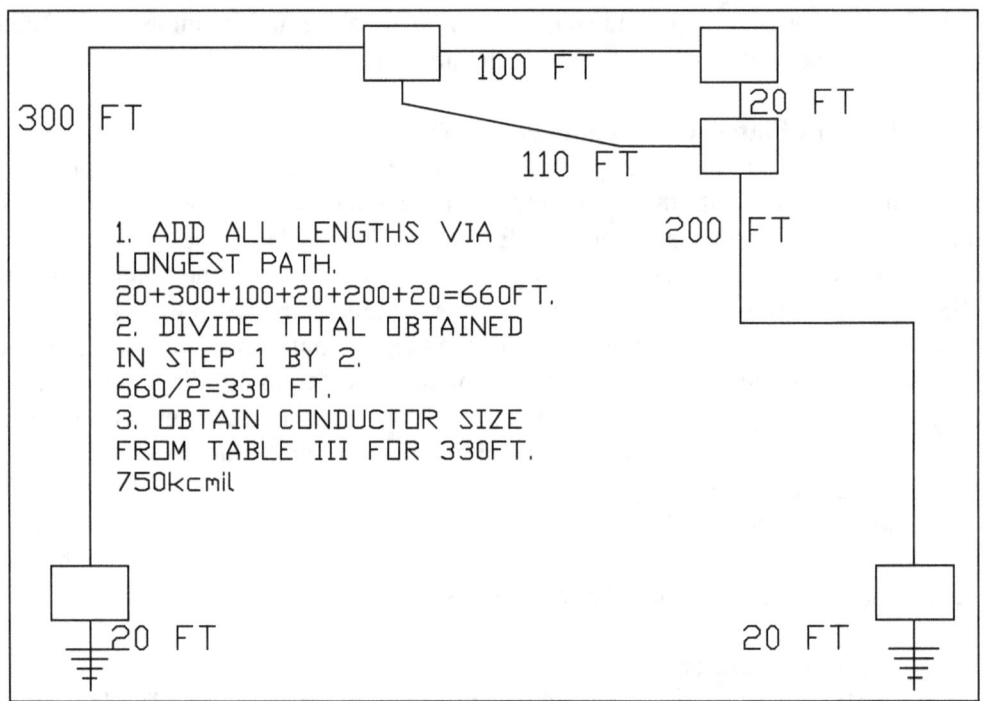

Figure 7. Multipoint Ground Cable Size Determination

3.10.3.3 Ground Conductors (Plate and Bus to Equipment).

Conductors from plates and buses in the multipoint system to equipment chassis shall be sized in accordance with Table III based on the maximum path length from the plate or bus to the equipment. . These conductors shall be color coded green with a bright orange tracer or shall be clearly marked for 4 in. at each end and wherever exposed with a green tape overlaid with a bright orange tracer. Where routed through raceways or wireways, the color-coding shall be visible by opening any cover. Where conductors are routed through cable trays, color-coding 4 in. long shall be provided at intervals not exceeding 3 ft. The use of uninsulated (bare) conductors are not permitted inside structures, except when used for short grounding jumpers, bonding jumpers, and similar items, that are not enclosed in conduit or raceway.

3.10.3.4 Protection

Provide mechanical protection for all cables in the electronic multipoint ground system where they may be subject to damage. This protection may be provided by conduit, floor trenches, routing behind permanent structural members, or other means as applicable. Where routed through metal conduit, the conduit shall be bonded to the cable at each end.

Table III. Size of Electronic Multipoint Ground Interconnecting Conductors

Cable Size	Max. Path Length		Bus Bar Size		Max. Path Length	
	Ft.	(m)	Inch	(mm)	Ft.	(m)
750 kcmil*	375	(114.3)	4 x 1/4	(100 x 6.4)	636	(193.9)
600 kcmil*	300	(91.4)	4 x 1/8	(100 x 3.2)	318	(96.9)
500 kcmil	250	(76.2)	3 x 1/4	(75 x 6.4)	476	(145.1)
350 kcmil	175	(53.3)	3 x 1/8	(75 x 3.2)	238	(72.5)
300 kcmil	150	(45.7)	2 x 1/4	(50 x 6.4)	318	(96.9)
250 kcmil	125	(38.1)	2 x 1/8	(50 x 3.2)	159	(48.5)
4/0 AWG	105	(32.0)	2 x 1/16	(50 x 1.6)	79	(24.1)
3/0 AWG	84	(25.6)	1 x 1/4	(25 x 6.4)	159	(48.5)
2/0 AWG	66	(20.1)	1 x 1/8	(25 x 3.2)	79	(24.1)
1/0 AWG	53	(16.2)	1 x 1/16	(25 x 1.6)	39	(11.9)
1 AWG	41	(12.5)				
2 AWG	33	(10.1)				
4 AWG	21	(6.4)				
6 AWG	13	(4.0)				

NOTE: kcmil* - Where these cables are not available, parallel cables may be used such as three 250 kcmil cables in place of one 750 kcmil cable, or two 300 kcmil cables in place of one 600 kcmil cable. The cable sizing is based on providing a cross-sectional area of 2000 circular mils per linear foot. The bus bar sizes are based on providing a cross-sectional area of 2000 cmil per linear foot.

3.10.3.5 Conductor Labeling

All multipoint ground system conductors shall be identified at each end. The source end shall indicate the point of utilization. The point of utilization termination shall be labeled indicating the source of the conductor. This shall be accomplished by either shrink embossed label or by a tie on tag. These conductors shall also be tagged every 50 ft. and in junction boxes in the manner above.

3.10.4 Building Structural Steel

All major structural members such as building columns, wall frames, roof trusses of steel frame buildings and other metal structures shall be made electrically continuous by bonding each major joint and interconnection in accordance with paragraph 3.14. The structural steel shall be connected to the EES as specified in paragraph 3.8.6. See also Figure 6

3.10.4.1 Metal Building Elements

The requirements of this paragraph apply to facilities which have sensitive receiver or computing systems and are located in areas where radiation from radar or other high power transmitters is expected. Metal building elements and attachments such as walls, roofs, floors, door and window frames, gratings and other architectural features shall be directly bonded to structural steel in accordance with paragraph 3.14. Where direct bonding is not practical, indirect bonds with copper cable conforming to Table III shall be used. Removable or adjustable parts and objects shall be grounded with an appropriate type bond strap as specified in paragraph 3.14.3. All bonds shall conform to the requirements of paragraph 3.14. Metal elements with a maximum dimension of 3 ft. (0.9 m) or less are exempt from the requirements of this paragraph.

3.10.5   Interior Metallic Piping Systems

The interior metallic cold water piping system shall be bonded in accordance with the NEC. A bond shall be required at the upper level of tower cabs. Where there is a separately derived power source for the tower cab, the interior metallic piping systems near the top of the ATCT, shall be referenced to the ground plate at the subjunction level.

3.10.5.1 Ground Connections

UL listed pressure clamps shall be used to bond pipes and tubes to the equipment ground system. In highly humid or corrosive atmospheres, adequate protection against corrosion shall be provided in accordance with paragraph 3.14.9. Do not use building main incoming water pipe as sole earth electrode reference. Many buried metal water pipes are coated with (insulating) corrosion-resistant fibers or are connected to buried plastic pipe, making the main incoming water pipe unsuitable as a primary earth electrode.

3.10.6   Electrical Supporting Structures

All metallic electrical support structures shall be electrically continuous and shall be directly bonded to the electronic multipoint ground system and to the EES.

3.10.6.1 Conduit

All metal conduit used for electronic signal and control wiring shall be grounded as follows:

(a) Conduit shall have a means to be bonded, prior to entering a structure, to a ground plate or bulkhead plate located outside the structure or directly to the EES. Plate(s) shall be bonded to the EES with an insulated 4/0 copper cable. Flexible conduits shall have a bonding jumper installed on the outside of the conduit.

(b) All joints between conduit sections and between conduit, fittings, and boxes shall be electrically continuous. All pipe and locknut threads shall be treated with a conductive lubricant rated for the metal prior to assembly. Surfaces shall be prepared in accordance with paragraph 3.14.8. Joints that are not otherwise electrically continuous shall be bonded with short jumpers of #6 AWG or larger copper wire. The jumpers shall be welded or brazed in place or shall be attached with clamps, split bolts, grounding bushings, or other devices approved for the purpose. All bonds shall be protected against corrosion in accordance with paragraph 3.14.9.4.

(c) Cover plates of conduit fittings, pull boxes, junction boxes, and outlet boxes shall be grounded by securely tightening all available screws.

(d) Every component of metallic conduit runs such as individual sections, couplings, line fittings, pull boxes, junction boxes and outlet boxes shall be bonded, either directly or indirectly, to the electronic multipoint ground system or facility steel at intervals not exceeding 50 ft. (15 m).

(e) Conduit brackets and hangers shall be securely bonded to the conduit and to the metal structure to which they are attached.

3.10.6.2 Cable Trays and Wireways
The individual sections of all cable tray systems for electronic conductors shall be bonded together with a minimum #6 AWG insulated copper conductor and each support bracket or hanger shall be bonded to the cable trays, which they support. All bonds shall be in accordance with procedures and requirements specified in paragraph 3.14. All tray assemblies for electronic conductors shall be connected, either directly or indirectly, to the electronic multipoint ground system or properly grounded facility steel within 2 ft (0.6 m) of each end of the run and at intervals not exceeding 50 ft. (15m). The resistance of each of these connections shall not exceed 5 milliohms. The minimum size, bonding conductor for connection of a cable tray and wireway to the MPG shall be #2 AWG copper conductor.

3.10.7   Secure Facilities
In all areas of facilities required to maintain communications security, equipment and power systems shall be grounded in accordance with NACSIM-5203 and MIL HB232A.

3.10.8   Multipoint Grounding of Electronic Equipment
When permitted by circuit design requirements, all internal ground references shall be directly bonded to the chassis and the equipment case. Where mounted in a rack, cabinet or enclosure, the electronic equipment case shall be bonded to the racks, cabinet or enclosure in accordance with paragraph 3.13.1. The DC resistance between any two points within a chassis or electronic equipment cabinet serving as ground shall be less than 25 milliohms total or 2.5 milliohms per joint. Shields shall be provided as required for personnel protection and electromagnetic interference (EMI) reduction. . To prevent the possibility of problems due to resonance of a single bonding strap, two widely spaced straps of unequal length shall be used to connect the equipment to the Multipoint grounding bus in the equipment cabinet. Bonding shall be in accordance with the recommended practices as expressed in IEEE Std 1100-1999 paragraph 8.5.4.6.

3.10.8.1 Electronic Signal Return Path
The electronic signal return path shall be routed with the circuit conductor. For axial circuits, the shield serves this purpose. The electronic equipment case and electronic multipoint ground system shall not be used as a signal return conductor.

3.10.8.2 Shield Terminations of Axial and Other Cables

All connectors shall be of a type and design that provides a low impedance path from the signal line shield to the electronic equipment case. If the electronic signal reference plane must be isolated from the electronic equipment case, and if the shielding effectiveness of the case must not be degraded, a connector of a tri-axial design that properly grounds the outer cable shield to the case shall be used. Shields of axial cables and balanced transmission lines shall be terminated by peripherally grounding the shield to the electronic equipment case. Bonding of connectors shall be in accordance with paragraph 3.14.14. The use of pigtails to terminate high frequency line shields outside the electronic equipment case shall not be permitted. Axial shields and connector shells shall be grounded to the electronic multipoint ground system at junction boxes, patch panels, signal distribution boxes, and other interconnection points along the electronic signal path. See paragraph 3.15.3 for more information on conductor and cable shielding.

3.10.9 Electronic Equipment Containing both Low and High Frequency Circuits

If the low and high frequency circuits in electronic equipment are functionally independent, and if construction and layout will permit separate electronic signal references, the low frequency circuits may be grounded in accordance with paragraph 3.11. If the low frequency and high frequency circuits share a common electronic signal reference, both circuits shall be grounded in accordance with paragraph 3.10.8.

3.10.9.1 Input and Output Electronic Signals

Where the low frequency signal reference is separate from the high frequency signal reference, low frequency input and output signals shall conform to paragraphs 3.11.7.2 and 3.11.7.4. High frequency input and output signals shall conform to paragraph 3.10.8. Where a common signal reference is used, low frequency analog input and output signals shall be balanced with respect to the signal reference. Extreme care shall be taken to maintain isolation between the single point ground system and the electronic multipoint ground system, except at the main ground plate or EES.

3.11 Electronic Single Point Ground System Requirements

3.11.1 General

Electronic single point ground systems are not required in FAA facilities unless the equipment to be installed requires it. The retrofitting of a single point ground system is a feasible option in most facilities. The electronic single point ground system shall be isolated from the power grounding system, lightning protection system and electronic multipoint ground system (except at the main ground plate). The electronic single point ground system shall be terminated at the main ground plate or to the EES, whichever is the closest. The electronic single point ground system shall be configured to minimize cable lengths. Conductive loops shall be avoided by maintaining a trunk and branch arrangement as shown in Figure 8.

3.11.2 Ground Plates

Main, branch and feeder ground plates shall be of copper and at least 12 in. (305 mm) long, 6in. (152 mm) wide, and 1/4 in. (6.4 mm) thick. The plates shall be mounted on phenolic or other non-conductive material of sufficient cross section to rigidly support the plates after all cables are connected. Bolts or other devices used to secure the plates in place shall be insulated or shall be of a non-conducting material. The plates shall be mounted in a manner that provides ready accessibility for future inspection and maintenance.

3.11.3 Isolation

The minimum resistance between the electronic single point ground and the electronic multipoint ground systems shall be 10 megohms. The resistance shall be measured after the complete network is installed and before connection to the EES or to the electronic multipoint ground system at the main ground plate.

3.11.4 Resistance

The maximum resistance between any ground plate and any cable connected to the plate shall not be greater than 1 milliohm.

3.11.5 Ground Cable Size

The size of the main, trunk and feeder ground cables shall be as follows:

3.11.5.1 Main Ground Cable

The main ground cable shall be an insulated 500 kcmil copper cable not exceeding 50 feet in length. The main ground cable shall be connected to the EES by exothermic weld and to the main ground plate with a UL listed connector and in accordance with paragraph 3.14.2 and as illustrated in Figure 8.

3.11.5.2 Trunk and Branch Ground Cables

An insulated trunk ground cable shall be installed in each facility from the main ground plate to each of the branch plates as shown in Figure 8. Insulated copper branch ground cables shall be installed between feeder and branch ground plates. These cables shall be routed to provide the shortest practical path. These cables shall be 4/0 AWG insulated copper conductors with a green and yellow marking for systems where the farthest feeder plate in the system is no more than 400 feet from the EES via the cable runs. For longer runs, select a cable size based on providing a cross sectional area of 500 cmil per running foot of cable length but in no case smaller than 250 kcmil. Trunk ground cables shall be exothermically welded or connected with UL listed double bolted connectors to the ground plates in accordance with paragraph 3.14.2.4 and shall be mounted as shown on the facility drawings.

3.11.5.3 Electronic Equipment Ground Cables

The cable from the feeder ground plate (branch ground plate if there is no need for a feeder ground plate in the cable run) to the isolated terminal or bus on the electronic equipment shall also meet the 500 cmil per running foot requirement. The minimum size cable shall be an insulated #6 AWG for cable lengths not more than 50 ft. (from the last plate to cabinet bus or equipment chassis), and for runs over 50 ft. the cable size shall be increased accordingly.

3.11.5.4 Interconnections

All connections to the single point ground system shall be made on ground plates or buses. Split bolts, Burndy clamps and other connections to existing cables are not allowed.

3.11.6　　　　Labeling

The single point ground system shall be clearly labeled to preserve its integrity as described in the following sections.

3.11.6.1　Cable Identification

Trunk, branch and electronic equipment ground cables shall be color coded green with a bright yellow tracer. Where cables are concealed and not color coded, any exposed portion of the cable and each end of the cable for a minimum length of 2 ft. (0.6 m) shall be color coded by green tape overlaid with bright yellow tape to form the tracer. Where routed through raceways or wireways, color coding shall be visible by opening any cover. Where conductors are routed through cable trays, color coding 4 in. in length shall be applied at intervals not exceeding 3 ft. All single point ground system conductors shall be identified at each end. The source end shall indicate the point of utilization. The point of utilization termination shall be labeled indicating the source of the conductor. This shall be accomplished by either a shrink embossed label or by a tie on tag. These conductors where exposed shall also be tagged every 50 ft. and in junction boxes in the manner above.

3.11.6.2　Ground Plate Labeling

All ground plates shall be protected with a clear plastic protective cover spaced 3/4 in. (19 mm) from the plate and extending 1 in. (25.4 mm) beyond each edge. This cover shall have a green label with distinguishing bright yellow slashes attached bearing the caption: "CAUTION, ELECTRONIC SINGLE POINT GROUND" in black 3/8-in. high (10 mm) characters.

3.11.7　　　　Equipment Requiring Electronic Single Point Grounds

When electronic equipment performance dictates an isolated electronic single Point ground system for proper operation, all the equipment and its installation shall comply with the following:

3.11.7.1　Electronic Single Point Ground System

The single Point ground system or plane shall be isolated from the electronic equipment case. If a metal chassis is used as the electronic single point ground, the chassis shall be floated relative to the case. Design practices shall be such that the single point ground of the electronic equipment can be properly interfaced with other electronic equipment without compromising the system. If necessary, this single point ground system may be filtered for high frequencies.

3.11.7.2　Input and Output Signals

The "high" and "low" sides of input and output signals shall be isolated from the electronic equipment case and balanced with respect to the signal reference. Operating and adjusting controls, readouts or indicating devices, protective devices, monitoring jacks and signal connectors shall be designed to isolate both the high and low side of the signal from the case.

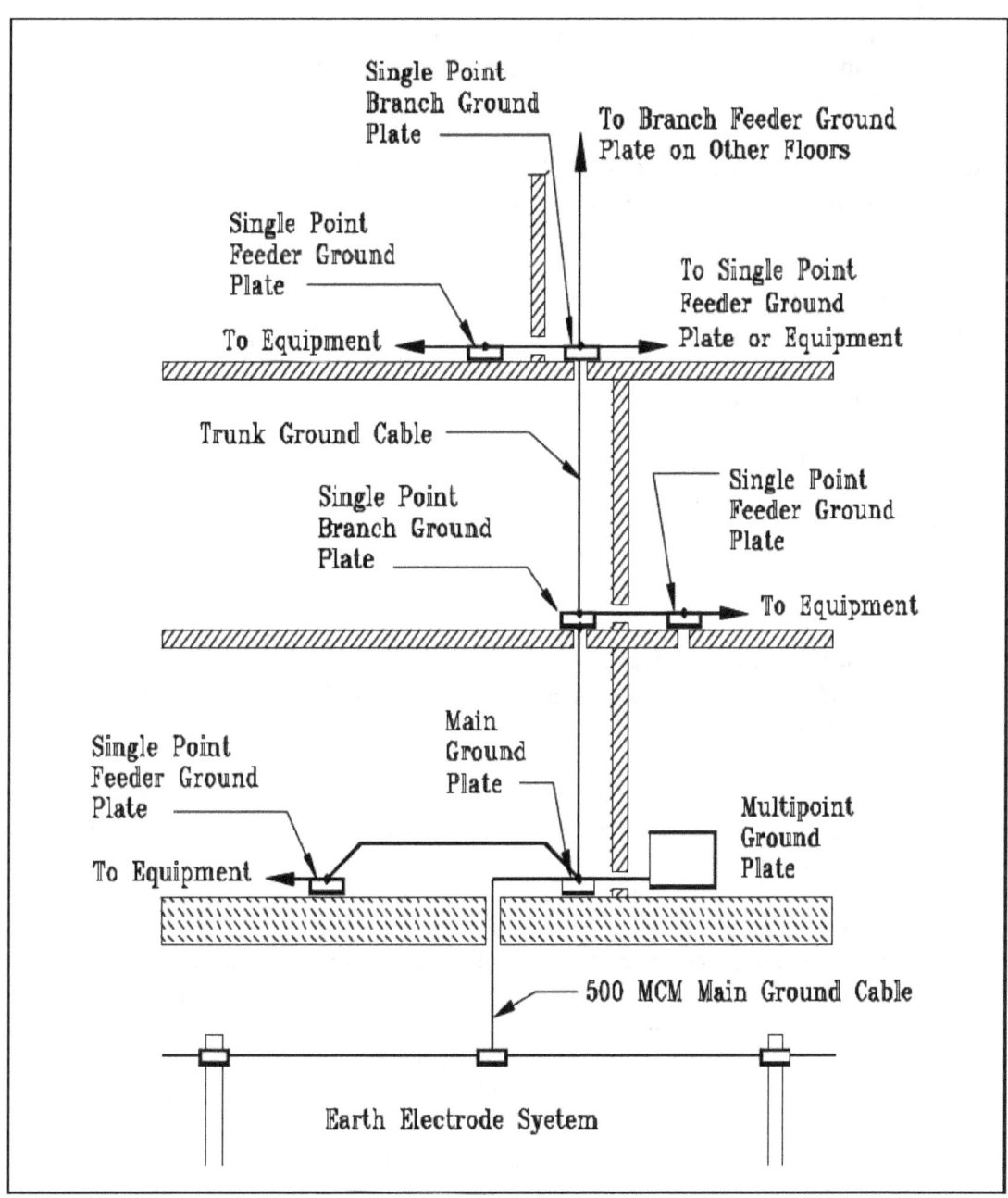

Figure 8. Electronic Single Point Ground System Installation

3.11.7.3 Electronic Single Point Ground Terminal(S)

Insulated single point ground system terminal(s) shall be provided on each electronic equipment case where an isolated signal reference is required. The single point ground reference for the internal circuits shall be connected to this terminal. This terminal(s) shall be used to terminate cable shields, and to connect the isolated signal ground of the electronic equipment to the single point ground system in the facility. A connector pin, a screw or pin on a terminal strip, an insulated stud, jack or feed through, or an insulated wire shall be an acceptable terminal so long as each terminal is clearly marked, labeled, or coded in a manner that does not interfere with its intended function. These marks, codes, or labels shall be permanently affixed and shall utilize green with yellow stripes. Wire insulation shall be green with a yellow tracer.

3.11.7.4 Isolation

With all external power, signal and control lines disconnected from the electronic equipment, isolation between the single point ground system terminals and the case shall not be less than 5 megohms.

3.11.7.5 Electronic Signal Lines and Cables

Electronic signal lines shall be twisted shielded pairs with the shield insulated. Cables consisting of multiple twisted pairs shall have the individual shields isolated from each other. Cables with an overall shield shall have the shield insulated.

3.11.7.6 Termination of Individual Shields

Termination of individual shields shall be in accordance with paragraph 3.15.3.2.

3.11.7.7 Termination of Overall Shields

Termination of overall shields shall be in accordance with paragraph 3.15.3.3.

3.11.7.8 Single Point Grounding of Electronic Equipment

Each single point ground terminal shall be connected to the facility single point ground system in accordance with the following:

(a) Individual units or pieces of electronic equipment which by nature of their location or function cannot or should not be mounted with other electronic equipment, shall have an insulated copper cable installed between the electronic single point ground terminal specified in paragraph 0 and the nearest electronic single point ground system ground plate. This cable shall have a cross-sectional area of 500 circular mils per linear foot.

(b) Where two or more units or pieces of electronic equipment are mounted together in a rack or cabinet, a single-ground bus bar shall be installed as shown in Figure 9. The bus bar shall be copper and shall provide a minimum cross-sectional area of 125,000 square mils. The bus bar shall be drilled and tapped for #10 screws. The holes shall be located as required by the relative location of the isolated electronic single point grounding terminals on the electronic equipment. The bus bar shall be mounted on insulating supports that provide at least 10 megohms DC resistance between the bus bar and the rack or cabinet.

(c) Each electronic equipment isolated single point ground terminal shall be interconnected to the bus bar by means of a solid or flexible tinned copper jumper of sufficient cross sectional area so that its resistance is 5 milliohms or less (#6 AWG minimum). The jumper shall be insulated or mounted in a manner that maintains the required degree of isolation between the reference conductor and the enclosure. The interconnecting jumper shall be attached to the bus bar at a point nearest to the single point ground terminal to which the strap is attached. An insulated copper cable shall be installed from the bus bar in the cabinet to the nearest electronic single point ground system. This cable shall provide at least 500 circular mils per linear foot, and must be a minimum of a #6 AWG.

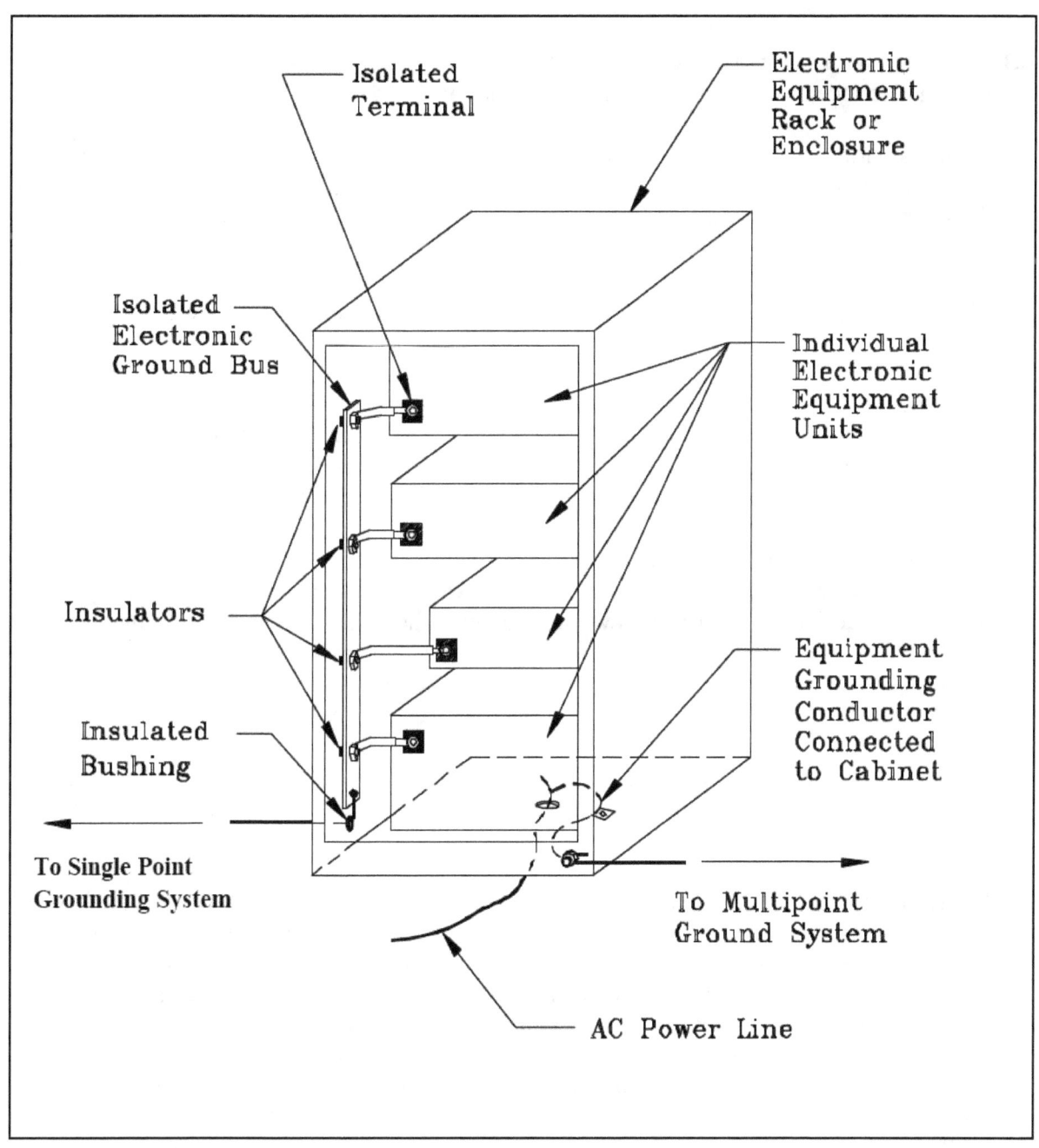

Figure 9. Single Point Electronic Ground Bus Bar Installation in Rack or Cabinet

3.12 National Electrical Code (NEC) Grounding Compliance

3.12.1 General

The facility electrical grounding shall exceed the requirements of NEC Article 250 as specified herein. The electronic multipoint ground system shall not replace the equipment grounding conductor required by the NEC.

3.12.2 Grounding Electrode Conductors

Grounding electrode conductors shall conform to the following:

(a) Premises wiring, required by the NEC to be grounded, shall have the neutral (grounded conductor) connected to the EES by a copper grounding electrode conductor at the service disconnecting means. The grounding electrode conductor shall be sized in accordance with the NEC, but in no case shall the wire size be smaller than #2 AWG.

(b) The grounding connection for services shall be made to the neutral bus in the service disconnecting means.

(c) The grounding electrode conductor connecting the neutral wire (grounded conductor) to the EES shall be continuous and unspliced, except where splices are permitted by the NEC. When a grounding electrode conductor is routed through a metal enclosure, e.g., conduit, the enclosure shall be bonded at each end to the grounding electrode conductor.

(d) Except as specified in sub-paragraph (e) for approach lighting systems with a constant voltage source used to supply flashers and convenience outlets the following applies. Where one facility receives its electrical power from another facility, the equipment grounding conductor shall be carried with the phase and neutral conductors in the same conduit or raceway and the grounded conductor (neutral) of the receiving facility shall not be connected to the equipment grounding conductor or grounding electrode at that facility.

(e) For approach lighting systems with a constant voltage source used to supply flashers and convenience outlets, the grounded conductor (neutral) shall be bonded to the counterpoise wire and to the grounding electrode at each light station.

(f) For separately derived systems, the grounding electrode conductor shall be connected from the grounded conductor in the first system disconnecting means directly to the nearest electrically continuous, effectively grounded structural steel. Where it is not feasible to connect the grounding electrode conductor to structural steel the EES may be used. This grounding electrode conductor shall be copper and sized in accordance with NEC requirements, except that this conductor shall not be smaller than #2 AWG. The bonding conductor between the equipment grounding conductor and the grounded conductor shall be installed in this disconnecting means. Equipment grounding conductors (safety grounds) shall be bonded to the grounded conductor in the first system disconnecting means. These equipment grounding conductors shall be green insulated, unspliced and the same size as the associated phase conductors. No neutral-to-ground connection shall be made at the load side disconnecting means enclosure.

3.12.3 Equipment Grounding Conductors

Each piece of electrical and electronic equipment shall be grounded. The equipment grounding conductor shall be a green insulated wire and run in the same raceway as its related phase and neutral conductors. Cord connected equipment shall include the equipment grounding conductor as an integral part of the power cord. Where power is supplied to electronic equipment through a cable and connector, the connector shall contain a pin to continue the equipment grounding conductor to the equipment chassis. Conduit or cable shields shall not be used as the equipment grounding conductor. All installations shall be in accordance with the NEC, FAA-C-1217 and with the following:

(a) See paragraph 3.13.2

(b) Equipment grounding conductors shall be the same size as the associated phase conductors.

(c) Grounding terminals in all receptacles on wire mold or plug mold strips shall be hardwired to an equipment grounding conductor. Strips that depend upon serrated or toothed fingers for grounding shall not be used.

(d) All flexible steel conduits shall contain an equipment grounding conductor. In addition to the internal equipment grounding conductor, an external bonding jumper shall be provided on all flexible metal conduits. This bonding jumper shall be a #6 AWG stranded copper conductor. The bonding jumper shall terminate on approved grounding fittings at each end of the flexible metal conduit.

(e) Individual equipment grounding conductors shall be installed in all branch circuits and feeders in parallel with the phase and grounded conductors.

3.12.4 Color Coding of Conductors

3.12.4.1 Grounded Conductors
Color coding of grounded conductors shall be consistent throughout the facility as follows:

(a) Neutral conductors (grounded conductors) shall be insulated and color coded white for 120/208V and 120/240 and natural gray for voltages above 120/240. Conductors larger than #6 AWG may be re-identified as the grounded (neutral) conductor **except** that green conductors shall not be re-identified. White or natural gray conductors shall be used in accordance with NFPA-70 (NEC) article 200.7

(b) In any room, conduit, pullbox, raceway, or cable tray, where two or more grounded conductors of different systems are present (branch circuits, feeders, services, voltages, etc.), the grounded conductors shall be clearly identified. The identification of the grounded conductors for each system shall be consistent throughout the facility. The grounded conductor of one system may be white, natural gray or re-identified. The grounded conductors of the other systems shall be identified by tape or by an identifiable colored stripe (not green) on white insulation.

(c) Color coding of grounded conductors shall be applied at each connection and at every point where the conductor is accessible. Where routed through raceways or wireways, the color coding shall be visible by removing or opening any cover. Where conductors are routed through cable trays, color coding 3 in. (75 mm) in length shall be provided at intervals not exceeding 3 ft. (0.9 m).

3.12.4.2 Equipment Grounding Conductors

Equipment grounding conductor color-coding shall be consistent throughout the facility as follows:

(a) Electrical equipment grounding conductors shall be solid green in color. Insulated conductors larger than #6 AWG may be re-identified with green tape. White or natural gray conductors shall not be re-identified as equipment grounding conductors. The equipment grounding conductor from the grounding terminal of the isolated ground pin receptacle to the service ground terminal shall be color coded green with yellow and red bands.

(b) Color-coding of equipment grounding conductors shall be applied at each connection and at every point where the conductor is accessible. Where routed through raceways or wireways, the coding shall be visible by removing or opening any cover. Where conductors are routed through cable trays, color coding 3 in. (75 mm) long shall be provided at intervals not exceeding 3 ft. (0.9 m).

3.12.4.3 Control and DC Power Cables

Color-coding for conductors in control cables shall be in accordance with NEMA Standard WC-5. DC power conductors, including battery cables, shall be color coded as follows: a red for positive conductor and black for a negative conductor. The red conductor shall be marked with a positive (+) symbol and the black conductor shall be marked with a (-) symbol.

3.12.5 Conductor Routing

The neutral (grounded conductor) and equipment grounding conductors shall be routed through the same raceway, or cable tray as the phase conductors. Power conductors shall not be routed in the same conduit or enclosed raceway with control, communications, or signal conductors or cables. Where power cables must be routed with electronic cables, the power conductors shall be twisted in accordance with Table IV. For conductor sizes larger than #2 AWG, the conductors shall be twisted to the maximum extent practical. The minimum separation distance between power and signal cables should be in accordance with paragraph 3.15.4. Power and signal cables should avoid crossing each other, but if it is necessary to have power and signal cables cross then it shall be at right angles.

3.12.6 Non-Current-Carrying Metal Equipment Enclosures

Metal enclosures shall meet the following requirements:

(a) All non-current-carrying metal enclosures such as conduit, raceways, wireways, cable trays and panel boards shall be electrically continuous. Insulating finishes shall be removed between grounding/bonding areas of mating surfaces or bonding jumpers. All rigid galvanized steel (ferrous rigid metal) conduits shall be equipped with grounding (bonding) bushings at each end and a bonding jumper the same size as the equipment grounding conductor as shown in Figure 10.

(b) Maximum use shall be made of ferrous materials for enclosures, conduits, raceways, and cable trays to provide shielding from magnetic fields (EMI and RFI).

(c) All battery supporting racks shall be bonded either directly to the EES or to any grounded structure with a #2AWG conductor.

Table IV. Minimum Number of Twists for Power Conductors

Size (AWG)	Twists per foot (0.3 m)			
	#of Conductors			
	Two	Three	Four	Five
12	7	5	4	3
10	6	4	3	2.5
8	5	4	3	2
6	4	3	2	1.5
4	3	2	1.5	1
2	2.5	2	1.5	1

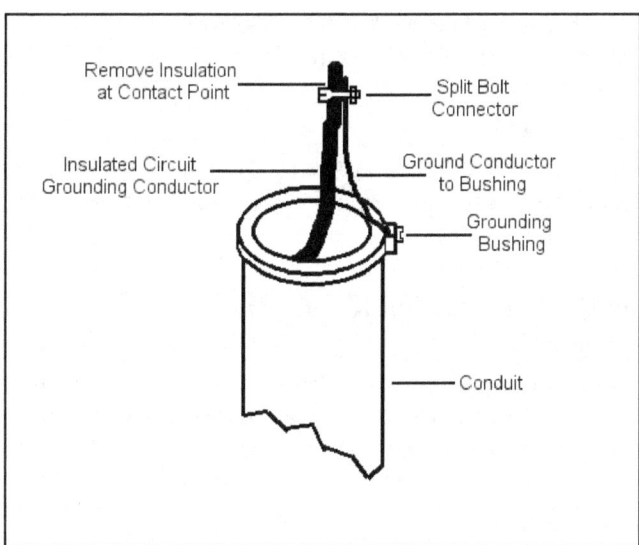

Figure 10. Bonding of Conduit and Grounding Conductor

3.13   Other Grounding Requirements

3.13.1   Electronic Cabinet, Rack, and Case Grounding

All electronic cabinets, racks, and cases shall provide a grounding terminal or means whereby a grounding jumper or wire can be mechanically connected through an electrically conductive surface to the basic frame. The metal enclosure of each individual unit or piece of electronic equipment shall be bonded to its cabinet, rack, or directly to the electronic multipoint ground system.

3.13.1.1 Mounting Surfaces
Electronic equipment mounting surfaces on cabinets and racks shall be free of non-conducting finishes. Mounting surfaces for electronic equipment that will be mounted in cabinets or racks shall also be free of non-conducting finishes.

3.13.2   Receptacles
Receptacles (convenience outlets) shall be provided with a ground terminal. An equipment grounding conductor whose path is electrically continuous and is in the same raceway or cable as the power conductors feeding the receptacles shall be connected to the ground terminal.

3.13.2.1   Isolated Grounding Receptacles
Where required for the possible reduction of electrical noise (electromagnetic interference) on the grounding circuit, receptacles shall be permitted in which the grounding terminal is purposely isolated from the receptacle mounting. The receptacle-grounding terminal shall be grounded by means of an isolated equipment-grounding conductor run with the circuit conductors (phase and neutral). This grounding conductor shall be permitted to pass through one or more panelboards without connection to the panelboard-grounding terminal. This equipment-grounding conductor shall terminate within the same building or structure directly at the equipment grounding conductor terminal of the applicable derived system or service and shall be color-coded green with yellow and red bands at each end and wherever it passes through a box. A second insulated equipment-grounding conductor shall be installed from the metal outlet box housing the receptacle to the panelboard feeding this receptacle. This conductor shall be bonded at one end to the metal outlet box and connected at the other end to the ground bus in the panelboard feeding the isolated ground pin receptacle. All equipment grounding conductors shall be run in the same conduit with their associated phase and neutral conductors. This equipment grounding conductor shall be color coded green.

3.13.3   Equipment Power Isolation Requirements
Prior to installation, the resistance from each conductor (including AC neutral) and the equipment case and any single point electronic ground connectors shall be measured. The isolation between these points shall measure 10 megohms or greater with the power switch in the on position.

3.13.4   Portable Equipment
Portable electrical or electronic equipment cases, enclosures, and housing shall be considered to be adequately grounded for fault protection through the equipment grounding conductor of the power cord, provided continuity is firmly established between the case, enclosure or housing, and the receptacle ground terminal. The power cord equipment grounding conductor shall not be used for signal grounding.

3.13.5   Fault Protection
Equipment parts such as panels, covers, knobs, switches, and connectors that are conductive and subject to human contact during operation and maintenance shall be prevented from becoming electrically energized when there are faults or component failures. Such parts shall be grounded by a low impedance path to the chassis or cases of the equipment on which they are mounted.

When grounded in accordance with paragraph 3.13.1, equipment chassis, cases, racks, cabinets, and other enclosures shall be considered adequately grounded for fault protection.

3.13.5.1 Metal Control Shafts
Metal control shafts shall be grounded to the equipment case through a low impedance path provided by close-fitting conductive gaskets, metal finger stock, or grounding nuts.

3.13.5.2 Shielded Compartments
Shields shall be bonded to the chassis for fault protection in accordance with paragraph 3.14.

3.13.6 AC Power Filters
All filter cases shall be directly bonded in accordance with paragraph 3.14.11 to the equipment case or enclosure. Filter leakage current shall not exceed 5 milliamperes (ma) per filter. Where practical and where the equipment is compatible, common power line filters shall be used to limit total leakage current. Transient suppression devices, components or circuits shall be installed in accordance with paragraph 3.6.

3.14 Bonding Requirements

3.14.1 Resistance
Unless otherwise specified in this standard, all bonds shall have a maximum DC resistance of 1 milliohm when measured between the bonded members with a 4-terminal milliohmmeter.

3.14.2 Methods
Bonding for electrical purposes shall be accomplished by a method that provides the required degree of mechanical strength, achieves and maintains the low value of low frequency and high frequency impedance required for proper functioning of the equipment, and is not subject to deterioration through vibration or corrosion in normal use. The surface contact area of bolted connections to flat surfaces in the lightning protection system shall be 3 square inches or greater. Soft soldered or brazed connections shall not be used for any part of the power grounding system, EES or the lightning protection system (air terminals, roof conductors, down conductors, fasteners, and conduit). Soft solder shall only be used to improve conductivity at joints already secured with mechanical fasteners. Soft solder shall not be used to provide mechanical restraint.

3.14.2.1 Exothermic Welds
Exothermic welds may be used for any type of bond connection specified herein. Exothermic welds shall be used for all underground connections between earth electrodes, counterpoise cable and other connections to the EES. Exothermic welds may not be possible between certain materials, shapes, or in hazardous locations, i.e., near fuel tanks, where nearby objects may be damaged, etc. In these cases, connections using UL listed connectors will be allowed.

3.14.2.2 Welds
Welds shall meet the following minimum requirements.

(a) Welds shall support the mechanical load demands on the bonded members.

(b) On members with a maximum dimension of 2 in. (50.8 mm) or less, the weld shall extend completely across the side or surface of the largest dimension.

(c) On members with a maximum dimension between 2 in. (50.8 mm) and 12 in. (305 mm), one weld of at least 2 in. in length shall be provided.

(d) On members with a dimension of 12 in. (305 mm) or more, two or more welds, each not less than 2 in. (50.8 mm) in length shall be provided at uniform spacing across the surface. The maximum spacing between welds shall not exceed 12 in.

(e) At butt joints, complete penetration welds shall be used on all members whose thickness is ¼ in. (6.4 mm) or less. Where the thickness of the members is greater, the depth of the weld shall be more than ¼ in.

(f) Fillet welds shall have an effective size equal to the thickness of the members.

(g) At lap joints between members whose thickness is less than ¼ in. (6.4 mm), double fillet welds shall be provided.

(h) For metal interfaces that are required to be RF-tight, the interface shall be continuously welded.

3.14.2.3 Dissimilar Metals

(a) Only exothermic welding shall be used for the permanent bonding of copper conductors to steel or other ferrous structural members. Where the combustion products of a standard exothermic weld may present problems, a smokeless exothermic process is commercially available. Exothermic welds may not be possible between certain materials, shapes, or in hazardous locations, i.e., near fuel tanks, where nearby objects may be damaged, etc. In these cases, connections using UL listed connectors will be allowed.

(b) All residual fluxes shall be removed or neutralized to prevent corrosion.

(c) Brazing material shall meet the requirements for dissimilar metals as specified in, Table V

Table V. Acceptable Couplings Between Dissimilar Metals

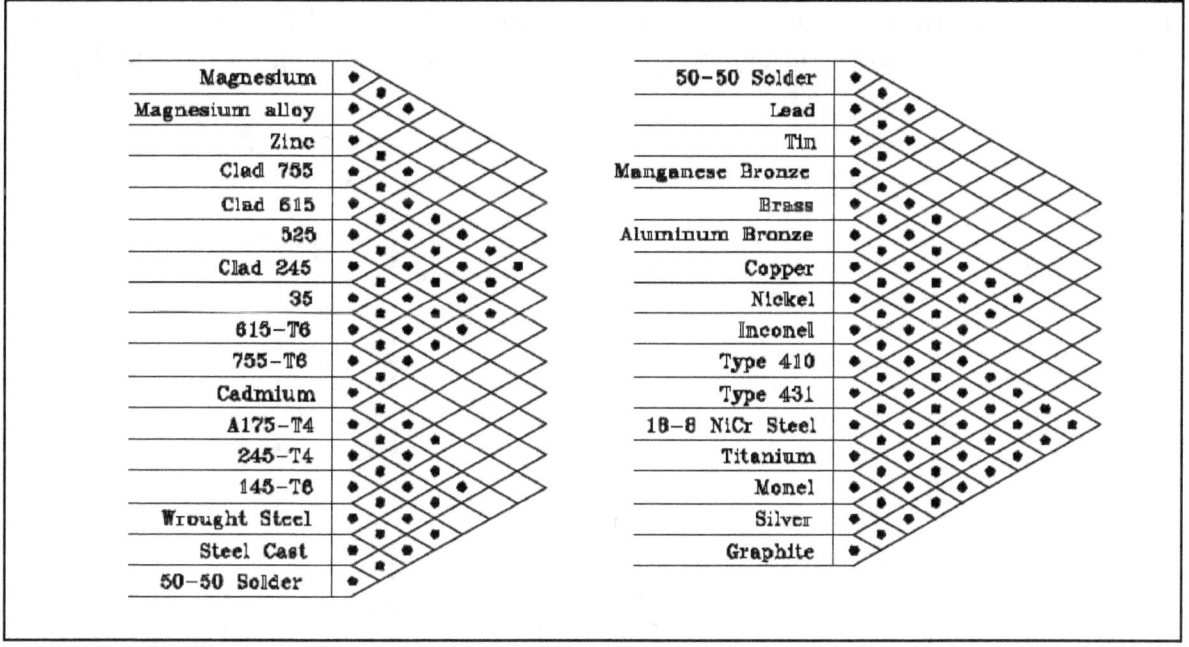

Notes:
1. Stainless steels, nickel, and inconel are considered passive on this chart.
2. Each metal on the chart is considered anodic (sacrificial) to the metals following it.
3. A solid dot (•) indicates an acceptable combination.

3.14.2.4 Mechanical Connections

3.14.2.4.1 Bolted Connections

Bolts and machine screws shall be used primarily as mechanical fasteners for holding the component members of the bond in place. They shall be tightened sufficiently to maintain the contact pressures required for effective bonding but shall not be over-tightened to the extent that deformation of bond members occurs. Disc springs (Belleville spring washers) shall be installed to prevent loosening. Bolts and screws other than those intentionally used to attach bonding straps or conductors, shall not be used in lieu of dedicated bonding jumpers

(a) All bolted connections shall conform to the torque requirements in Table VI. All bolted connections shall utilize SAE Grade 5 stainless steel bolts and nuts.

(b) Bolted connections shall be assembled in the order shown in Figure 11. Additional load distribution washers, if used, shall be positioned directly underneath the bolt head. Disc springs shall be between the nut and the load distribution washer. Washers shall not be placed between bonded members. (See paragraph 3.14.8 for surface preparation.) Load distribution washers shall comply with ANSI B18.22.1 for Stainless steel washers, Wide Series, Type B. **Note:** Table VI provides sample part numbers for one manufacturer; other manufacturers of disc spring washers may be equally suitable.

3.14.2.4.2 Sheet Metal Screws

Sheet metal screws shall not be used to provide a continuous and permanent electrical bond. The use of sheet metal screws shall be restricted to the fastening of covers. These covers are to eliminate dust or other foreign matter from the equipment, and to discourage unauthorized or untrained personnel access to the equipment.

Table VI. Torque Requirements for Bolted Bonds

Bolt Size	Torque (ft-lbs)	Bolt Area Load (psi)	Washers Required	Solon Part Number*
#8	3	1200	3	08-H-35-177
#10	5	1500	2	010-M-50-177
1/4 in.	10	2500	3	4-EH-70-301
5/16 in.	21	4000	3	5-EH-80-301
3/8 in.	34	5500	3	6-EH-89-301
7/16 in.	55	7500	6	7-L-70-301
1/2 in.	83	10,000	2	8-18-125-301
9/16 in.	117	12,500	N/A	N/A
5/8 in.	167	16,000	3	10-EH-150-177
3/4 in.	288	23,000	3	12-EH-168-177
7/8 in.	452	31,000	3	14-EH-168-177
1 in.	567	40,000	3	15-H-187-177

*Other manufacturers of disc spring washers may be equally suitable

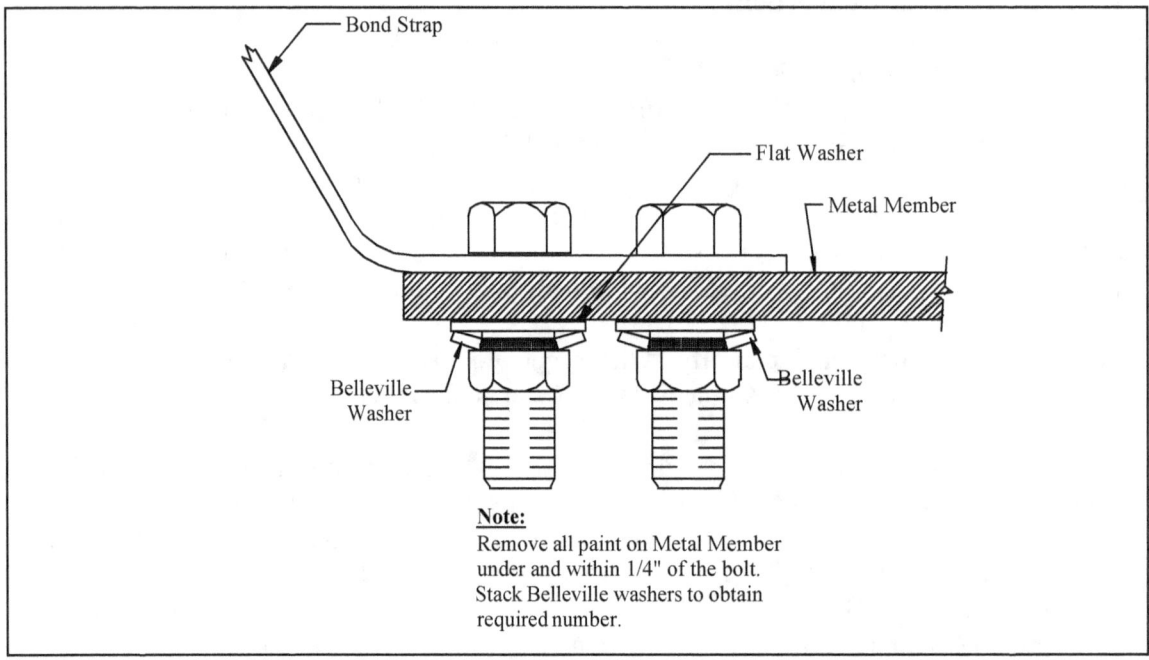

Figure 11. Order of Assembly for Bolted Connections

3.14.2.4.3 Hydraulically Crimped Terminations

Mechanical connections such as a Burndy "Hyground Connector", Thomas and Betts Compression Connector or approved equivalent, when operated at the manufacturer's recommended pressure to develop a minimum force of 12 tons, are acceptable as FAA approved pressure connectors. (These connectors are not acceptable in the lightning protection system.)

3.14.2.4.4 Soldering

Mechanical bonds may be improved by the use of silver solder to exclude contaminants from the mating surfaces. Soldered mechanical bonds shall be carefully made prior to applying solder to prevent cold solder joints. Soldered mechanical connections shall not be used for any part of the power grounding system or the lightning protection system. See FAA Order 6950.20, Chapter 5, paragraph 99 for additional information.

3.14.2.4.5 Riveting

Rivets shall be employed primarily as mechanical fasteners to hold multiple smooth, clean metal surfaces together or to provide a mechanical load bearing capability to a soldered bond. Rivets shall not be used as bonds for high frequency signals.

3.14.3 Bonding Straps and Jumpers

Bonding straps, including jumpers, shall conform to the following:

(a) Bonding straps shall be attached to the basic member rather than through any adjacent parts.

(b) Bonding straps shall be installed to be unaffected electrically by motion or vibration.

(c) Braided bonding straps shall not be used for bonding transmitters or other sources of RF fields.

(d) Bonding straps shall be installed whenever possible in areas accessible for maintenance and inspection.

(e) Bonding straps shall be installed so they will not restrict movement of the members being bonded or other members nearby which must be able to move as part of normal functional operation.

(f) Two or more bonding straps shall not be connected in series to provide a single bonding path.

(g) The method of installation and point of attachment of bonding straps shall not weaken the members to which they are attached.

(h) Bonding straps shall not be compression-fastened through non-metallic material.

(i) Bonding straps shall be designed not to have resonant impedances at equipment operating frequencies. Two short, low-impedance grounding straps between the multipoint grounding system and two corners of the equipment should be used. These straps should be connected as far apart as possible on the equipment (ideally on opposite corners) in order to reduce mutual inductance and they should have few bends or sags. Two straps with a 20% to 30% difference in length should be used so that if one strap experiences resonance, limiting current flow, the other strap will not.

(j) The length of the equipment bonding and grounding wire connections should be limited to 1/20th of a wavelength of the signal frequency, or about six inches at 100 MHz. In practice, this may not be possible so the length should be as short as possible.

(k) Broad flat conductors, with a large surface area (at least one inch wide) should be used for equipment grounding straps since they have a lower inductance than other (round) conductors. Care must be taken to insure that terminations of braids or straps be constructed in a manner which maintains their contact width. Reduction of contact width by utilizing standard round conductor terminal lugs is unacceptable.

3.14.4   Fasteners
Fastener materials for bonding aluminum and copper jumpers to structures shall conform to the materials listed in Table V. Acceptable Couplings Between Dissimilar Metals.

3.14.4.1   Counterpoise Cables
Counterpoise cables shall be attached to ground rods in accordance with the requirements of paragraph 3.8.6.

3.14.4.2   Underground Metallic Pipes and Tanks
Underground metallic pipes and tanks shall be bonded to the EES in accordance with the requirements of paragraph 3.8.6.

3.14.4.3   Steel Frame Buildings
Structural members (columns, wall frames, and roof trusses) shall be electrically continuous. Where joints are not electrically continuous, they shall be bridged to obtain continuity, with an exothermically welded #4/0 AWG stranded copper cable.

3.14.4.4   Interior Metallic Pipes
Interior metallic pipes and conduits shall be bonded in accordance with paragraph 3.10.5.

3.14.4.5   Electrical Supporting Structures
Conduit and cable trays shall be bonded in accordance with paragraph 3.10.6.

3.14.4.6   Flat Bars
Flat bars shall be bonded by minimum, SAE grade 5, high compression bolts.

3.14.5   Temporary Bonds
Alligator clips and other spring loaded clamps shall be employed only as temporary bonds while performing repair work on equipment or facility wiring.

3.14.6   Inaccessible Locations
All bonds in permanently concealed or inaccessible locations shall be exothermically welded.

3.14.7 Coupling of Dissimilar Metals
Compression bonding with bolts and clamps shall be used only between metals having acceptable coupling values as shown in Table V. When the base metals form couples that are not allowed, the metals shall be coated, plated, or otherwise protected with a conductive finish, or a washer made of a material compatible with each shall be inserted between the two base metals. The washer shall be constructed of passivated stainless steel. MIL-STD-889 provides specific information in this area.

3.14.8 Surface Preparation
All surfaces to be bonded shall be thoroughly cleaned to remove all dirt, grease, oxides, nonconductive films, or other foreign material. Paints and other organic coatings shall be removed by sanding or brushing down to the bare metal. The use of chemical removers shall be acceptable, provided that all residue is removed from the area to be bonded and provided that the chemical does not react with the base metal to produce nonconductive or corrosive products.

3.14.8.1 Paint Removal
Paints, primers, and other non-conductive finishes shall be removed from the metal base with appropriate chemical paint removers, or the surface shall be sanded with 500-grit abrasive paper or equivalent.

3.14.8.2 Inorganic Film Removal
Rust, oxides, and non-conductive surface finishes (anodized, galvanized, etc.) shall be removed by sand blasting, by using abrasive paper or cloth with 320-grit or finer, or by using an appropriate wire brush technique. Gentle and uniform pressure shall be employed when using abrasive papers or cloths or wire brushes to obtain a smooth, uniform surface. No more metal than necessary to achieve a clean surface shall be removed.

3.14.8.3 Area to Be Cleaned
All bonding surfaces shall be cleaned over an area that extends at least 1/4 in. (6.4 mm) beyond all sides of the bonded area on the larger member.

3.14.8.4 Final Cleaning
After initial cleaning with chemical paint removers or mechanical abrasives, the bare metal shall be wiped or brushed with an appropriate solvent meeting the requirements of P-D-680. Prior to bonding, surfaces not requiring the use of mechanical abrasives or chemical removers shall be cleaned with a dry cleaning solvent to remove grease, oil, corrosion preventives, dust, dirt, and moisture.

3.14.8.4.1 Clad Metals
Clad metal shall be carefully cleaned, to a bright, shiny, smooth surface, with fine steel wool or grit so the cladding material is not penetrated by the cleaning process. The cleaned area shall be wiped with dry cleaning solvent and allowed to air dry before completing the bond.

3.14.8.4.2 Aluminum Alloys
After cleaning of aluminum surfaces to a bright finish, a brush coating of alodine or other similar

conductive finish shall be applied to the mating surfaces.

3.14.8.5 Completion of the Bond
If an intentional protective coating is removed from the metal surface, the mating surfaces shall be joined within 2 hours after cleaning. If delays beyond two hours are necessary in corrosive environments, the cleaned surfaces must be protected with an appropriate coating that must be removed before completion of the bond.

3.14.8.6 Refinishing of Bond
Bonds shall be refinished so as to match the existing finish as close as possible within the requirements of paragraph 3.14.9.

3.14.8.7 Machined Surfaces
Where an RF-tight joint is necessary, both surfaces shall be machined smooth to provide uniform continuous contact through the joint area and an RF gasket (paragraph 0) shall be used to insure a low impedance path across the joint. Fasteners shall be positioned and distributed in a manner that maintains uniform pressure throughout the bond area.

3.14.8.8 Surface Platings or Treatments
Surface treatments that include platings provided for added wearability or corrosion protection shall offer high conductivity. Plating metals shall be electrochemically compatible with the base metals per paragraph 3.14.7. Unless suitably protected from the atmosphere, silver and other easily tarnished metals shall not be used to plate bond surfaces, except where an increase in surface contact resistance cannot be tolerated.

3.14.9 Bond Protection
All bonds shall be protected against weather, corrosive atmospheres, vibration and mechanical damage. Under dry conditions, a compatible corrosion preventive or sealant shall be applied within 24 hours of assembly of the bond materials. Under conditions exceeding 60% humidity, sealing of the bond shall be accomplished within 1 hour of joining.

3.14.9.1 Paint
If a paint finish is required on the final assembly, the bond shall be sealed with the recommended finish. Care shall be taken to assure that all means by which moisture or other contaminants may enter the bond are sealed. A waterproof type of paint or primer conforming to FAA-STD-012 shall be used if the recommended finish is not waterproof.

3.14.9.2 Inaccessible Locations
Bonds which are located in areas not reasonably accessible for maintenance shall be sealed with compatible permanent, waterproof compounds after assembly.

3.14.9.3 Compression Bonds in Protected Areas
Compression bonds between copper conductors or between compatible aluminum alloys located in readily accessible areas not exposed to weather, corrosive fumes, or excessive dust do not require sealing.

3.14.9.4 Corrosion Protection

All exterior and interior bonds exposed to moisture or high humidity shall be protected against corrosion. All interior bonds made between dissimilar metals shall be protected against corrosion in accordance with Table V and paragraph 3.14. All exothermic connections shall be cleaned of all residual slag. Protection shall be provided by a moisture proof paint conforming to the requirements of FAA-STD-012 or shall be sealed with a silicone or petroleum-based sealant to prevent moisture from reaching the bond area. Bonds protected by conductive finishes (alodine, iridite, et. al.) shall not require painting to meet the requirements of this standard.

3.14.9.5 Vibration

Bonds shall be protected from vibration-induced deterioration by assuring that bolts and screws employ lock washers, self-locking nuts, or jam nuts that are properly tightened and rivets that are securely seated.

3.14.10 Bonding Across Shock Mounts

Bonding straps installed across shock mounts or other suspension or support devices shall not impede the performance of the mounting device. They shall be capable of withstanding the anticipated motion and vibration requirements without suffering metal fatigue or other failures. Extra care shall be utilized in the attachment of bonding strap ends to prevent arcing or other forms of electrical noise generation from strap movement.

3.14.11 Enclosure Bonding

Subassemblies and equipment shall be directly bonded, whenever practical, at the areas of physical contact with the mounting surface.

3.14.12 Subassemblies

Subassemblies shall be bonded to the chassis utilizing the maximum possible contact area. All feed throughs, filters, and connectors shall be bonded around the periphery to the subassembly enclosure to maintain shield effectiveness. Covers shall exhibit intimate contact around their periphery, and contact shall be achieved and maintained through the use of closely spaced screws or bolts, or the use of resilient conductive gaskets, or both.

3.14.13 Equipment

The chassis or case of equipment shall be directly bonded to the rack, frame, or cabinet in which it is mounted. Flange surfaces and the contact surface on the supporting element shall be cleaned of all paint or other insulating substances in accordance with the requirements of paragraph 3.14.8. Fasteners shall maintain sufficient pressure to assure adequate surface contact to meet the bond resistance requirements in paragraph 3.14.1. Tinnerman nuts and sheet metal screws shall not be used for fasteners. If equipment must remain operational when partially or completely withdrawn from its mounted position, the bond shall be maintained by a moving area of contact or by the use of a flexible bonding strap. Except when necessary to maintain bonding during adjustments, maintenance, or when other constraints prevent direct bonding, the use of straps shall be avoided. Mechanical designs shall emphasize direct bonding.

3.14.14 Connector Mounting

All connectors shall be mounted so that intimate metallic contact is maintained between the connector and the panel to which it is mounted. Bonding shall be accomplished completely around the periphery of the flange of the connector. Both the flange surface and the mating area on the panel shall be cleaned in accordance with paragraph 3.14.8. All nonconductive material shall be removed from the panel as illustrated in Figure 12. After mounting of the connector, the exposed area of the panel shall be repainted or otherwise protected from corrosion in accordance with paragraph 3.14.9.

3.14.15   Shield Terminations
Cable shields shall be terminated in the manner specified by paragraphs 3.15.3.2, 3.15.3.3 and 3.10.8.2. Shields of axial cables shall be fastened tightly to the cable connector shell with a compression fitting or soldered connection. The cable shall be able to withstand the anticipated use without becoming noisy or suffering a degradation in shielding efficiency. Axial connectors shall be of a material that is corrosion resistant in keeping with requirements of FAA-G-2100. Low frequency shields shall be soldered in place or, if solderless terminals are used, the compressed fitting shall afford maximum contact between the shield and the terminal sleeve. Shield pigtails shall extend less than 1 inch from the point of breakaway from the center conductors of the cable.

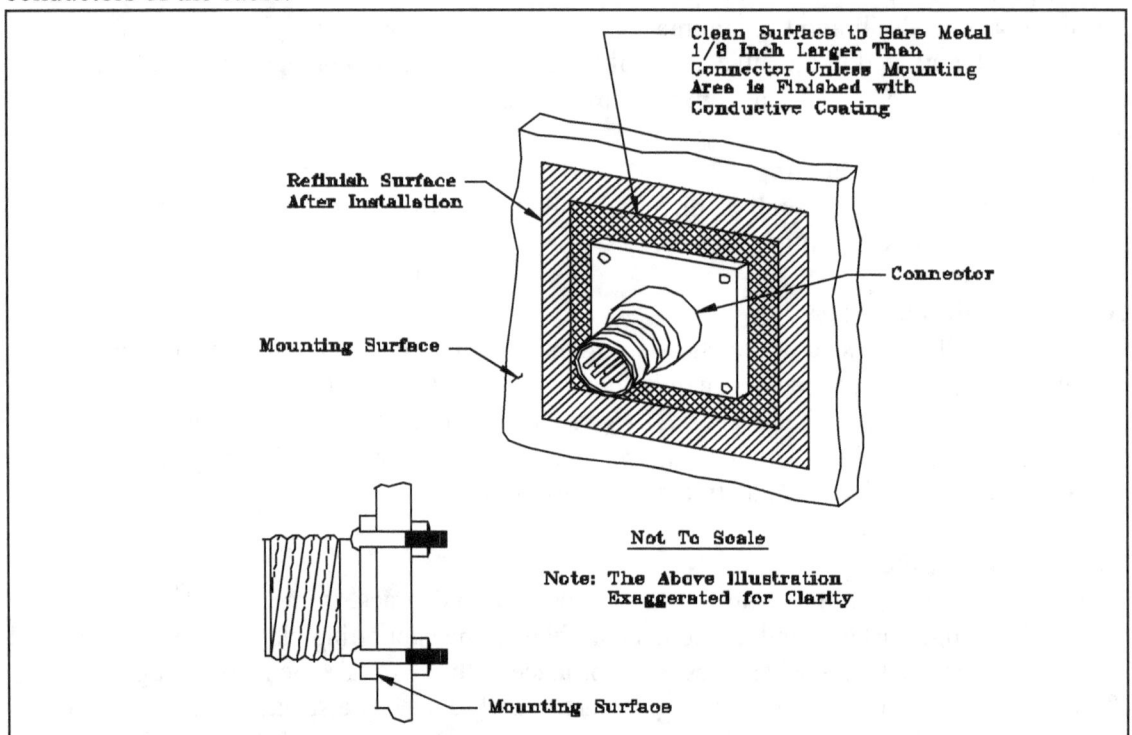

Figure 12. Bonding of Connectors to Mounting Surface

3.14.16   RF Gaskets
Conductive gaskets shall be made of corrosion resistant material, shall offer sufficient conductivity to meet the resistance requirements of paragraph 3.14.1, and shall possess adequate strength, resiliency, and hardness to maintain the shielding effectiveness of the bond. The surfaces of contact with the gasket shall be smooth and free of insulating films, corrosion, moisture, and paint. The gasket shall be firmly affixed to one of the bond surfaces by screws,

conductive cement, or other means that do not interfere with the effectiveness of the gasket; or a milled slot shall be provided that prevents lateral movement or dislodging of the gasket when the bond is disassembled. Gaskets shall be a minimum of 1/8 inch wide and of a reusable type. The gasket as well as the contact surfaces shall be protected from corrosion.

3.15 Shielding Requirements

3.15.1 Design
The facility design and construction shall incorporate both protective shields to attenuate radiated signals; and separation of equipment and conductors to minimize the coupling of interference. The equipment design shall incorporate component compartments and overall shields as necessary to meet the electromagnetic susceptibility and emission requirements of MIL-STD-461 as required by NAS-SS-1000 and FAA-G-2100. In addition, the design shall provide the shields necessary to protect personnel from hazardous voltages, high level electromagnetic fields, and x-rays that may be generated by equipment.

3.15.2 Facility Shielding
The shielding of facility buildings, shelters or equipment spaces shall be provided when other facility or environmental sources of radiation are of sufficient magnitude to degrade the operation and performance of electronic equipment. Unless otherwise specified, the bonding and grounding of metal structural components, building elements and the space separation of equipment and conductors shall be as indicated herein. Also, where rebar or a Faraday cage exists, it shall be connected to the EES with an exothermically welded #2 AWG copper conductor.

3.15.3 Conductor and Cable Shielding
Conductor and cable shielding shall comply with the following:

3.15.3.1 Signal Lines and Cables
Cables consisting of multiple twisted pairs shall have the individual shields isolated from each other. Cables with an overall shield shall have the shield insulated.

3.15.3.2 Termination of Individual Shields
Shields of pairs of conductors and the shield of cables containing unshielded conductors shall be terminated in accordance with the following:

(a) Shields shall be terminated as applicable for equipment operation.

(b) Shield terminations shall employ minimum length pigtails between the shield and the connection to the bonding halo or ferrule ring and between the halo or ferrule ring and the shield pin on the connector. The unshielded length of a signal line shall not exceed 1 in. (25 mm) with not more than 1/2 in. (13 mm) of exposed length as the desired goal.

(c) Shields, individually and collectively, shall be isolated from overall shields of cable bundles and from electronic equipment cases, racks, cabinet, junction boxes, conduit, cable trays, and elements of the electronic multipoint ground system. Except for one interconnection, individual shields shall be isolated from each other. This isolation shall be maintained in junction boxes, patch panels and distribution boxes throughout the cable run. When a signal line is interrupted such as in a junction box, the shield shall be carried through. The length of unshielded conductors shall not exceed 1 in. (25 mm). To meet this requirement, the length of shield pigtail may be longer than 1 in. but shall be the minimum required.

(d) Circuits and chassis shall be designed to minimize the distance from the connector or terminal strip to the point of attachment of the shield-grounding` conductor to the electronic signal reference. The size of the wire used to extend the shield to the circuit reference shall be as large as practical but shall not be less than #16 AWG or the maximum wire size that will fit the connector pin. A common shield ground wire for input and output signals, for both high level and low level signals, for signal lines and power conductors, or for electronic signal lines and control lines is prohibited.

(e) Nothing in this requirement shall preclude the extension of the shields through the connector or past the terminal strip to individual circuits or chassis if required to minimize unwanted coupling inside the electronic equipment. Where extensions of this type are necessary, overall cable or bundle shields grounded in accordance with paragraph 3.15.3.3 shall be provided.

3.15.3.3 Termination of Overall Shields

Cables that have an overall shield over individually shielded pairs shall have the overall shield grounded at each end either directly or through an SPD and at intermediate points in accordance with the following.

(a) Cable shields terminated to connectors shall be bonded to the connector shell as shown in Figure 13a or Figure 13b. The shield shall be carefully cleaned to remove dirt, moisture, and corrosion products. The connector securing clamp shall be carefully tightened to assure that a low resistance bond to the connector shell is achieved completely around the circumference of the cable shield. The bond shall be protected against corrosion in accordance with paragraph 3.14.9. The panel-mounted part of the connector shall be bonded to the mounting surface in accordance with paragraph 3.14.14.

(b) Where the cable continuity is interrupted such as in a junction box, the shield shall be carried through and grounded at the box. The length of unshielded conductors shall not exceed 1 in. (25 mm). To meet this requirement the length of shield pigtail may be longer than 1 inch, if necessary, to reach ground but shall be kept to a minimum.

(c) Cables which penetrate walls or panels of cases or enclosures without the use of connectors shall have their shields bonded to the penetrated surface in the manner shown in Figure 13c). Overall shields shall be terminated to the outer surface of cases to the maximum extent possible.

(d) Grounding of overall shields to terminal strips shall be as shown in Figure 14, Grounding Overall Shield to Terminal Strip.

(e) The shields of the individual pairs shall be grounded as specified in paragraph 3.15.3.2.

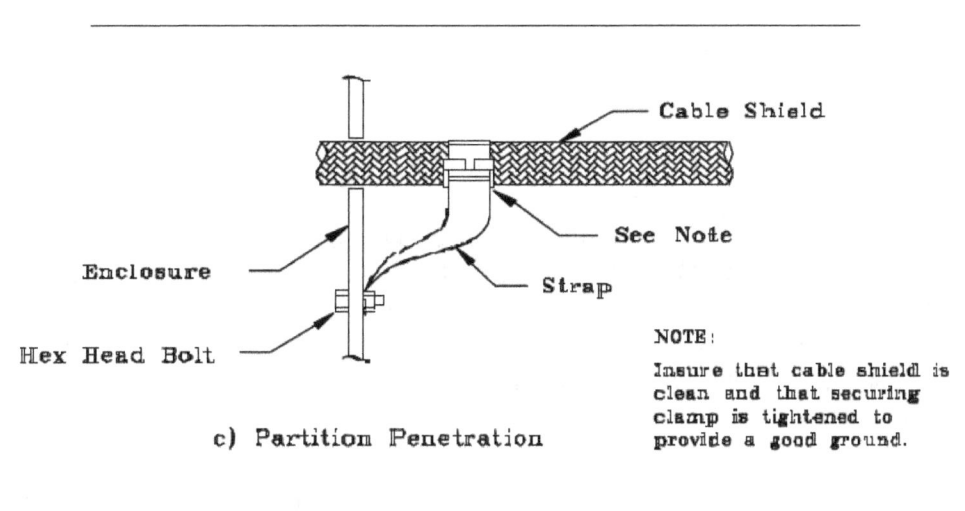

Figure 13. Grounding of Overall Cable Shields to Connectors and Penetrating Walls

3.15.4 Space Separation

The design and layout of facilities shall physically separate electronic equipment and conductors that produce interference from equipment and conductors that are susceptible to interference. In general, electronic equipment and conductors that carry, produce or use high levels of current (greater than 100 ma) or voltage (12V or more), including pulse power, can produce interference. Electronic equipment and conductors that carry, produce or receive low voltage or power levels are susceptible to interference. The minimum separation distance between power and signal cables shall be in accordance with Table VII.

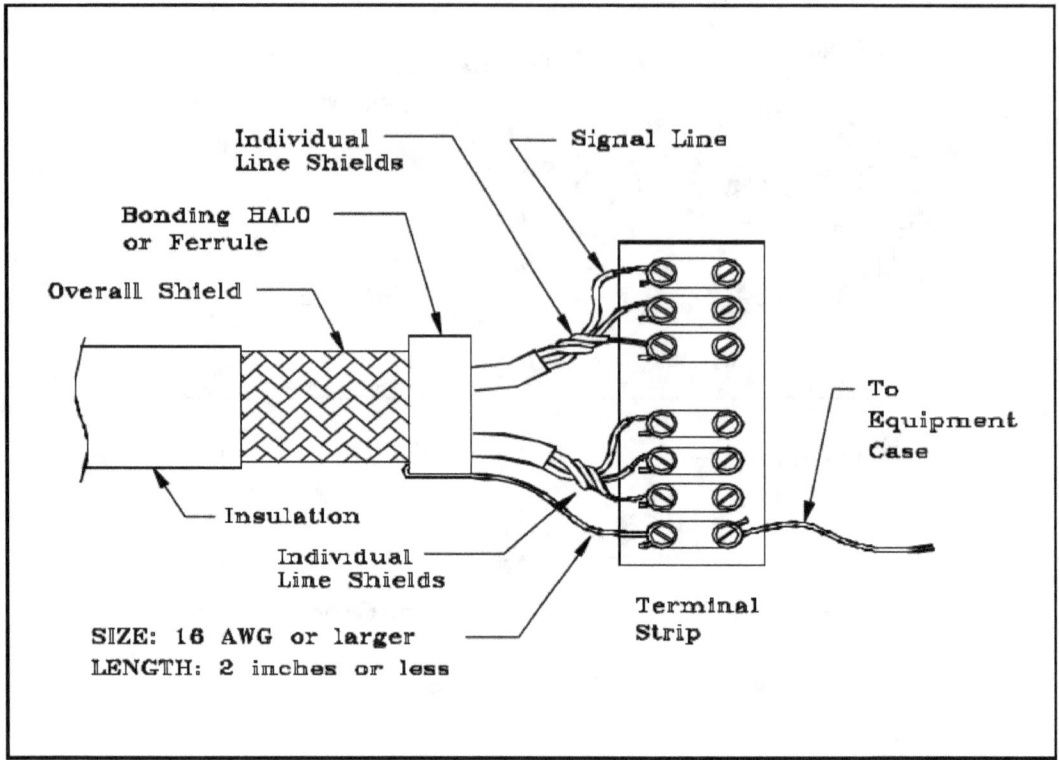

Figure 14. Grounding of Overall Cable Shield to Terminal Strip

3.15.5　Electromagnetic Environment Control
Shielding shall be integrated with other basic interference control measures such as filtering, wire routing, cable and circuit layout, signal processing, spectrum control, and frequency assignment to achieve the highest operational reliability of the equipment. Implementation procedures necessary to achieve the required filtering and shielding shall be detailed in the control plan described in paragraph 4.2 to include material requirements, shield configurations, placement and installation limitations, gasket utilization, filter integration, aperture control, bonding and grounding requirements, and wire routing and circuit layout constraints.

Table VII. Minimum Separation Distance Between Signal and Power Cables.

	Minimum Separation Distance		
Condition	< 2 kVA	2-5 kVA	> 5 kVA
Unshielded power lines or electrical equipment in proximity to signal conductors in open or nonmetal pathways.	5 in. (127 mm)	12 in. (305 mm)	24 in. (610 mm)
Unshielded power lines or electrical equipment in proximity to signal conductors in a grounded metal conduit pathway.	2.5 in. (64 mm)	6 in. (152 mm)	12 in. (305 mm)

| Power lines enclosed in a grounded metal conduit (or equivalent shielding) in proximity to signal conductors in a grounded metal conduit pathway. | - | 3 in. (76 mm) | 6 in. (152 mm) |

3.15.5.1 Materials
Shields shall be constructed of a material that provides the required degree of signal suppression without incurring unnecessary expense and weight. In the choice of the material, the amplitude and frequency of the signals to be attenuated, the characteristics of the electromagnetic field of the signal (i.e., the signal being coupled via inductive, capacitive, or free space means), configuration and installation constraints, and corrosion properties shall be considered.

3.15.5.2 Gaskets
Conductive gaskets conforming to paragraph 0 shall be utilized at joints, seams, access covers, removable partitions, and other shield discontinuities to the extent necessary to provide interference-free operation of the equipment under normal use and environmental conditions. Finger stock used on doors, covers, or other closures subject to frequent openings shall be installed in a manner that permits easy cleaning and repair.

3.15.5.3 Filter Integration
Filters on power, control, and signal lines shall be installed in a manner that maintains the integrity of the shield. AC power filters shall be completely shielded with the filter case grounded in accordance with paragraph 3.13.6. Filters for control and signal lines shall be placed as close as possible to the point of penetration of the case to avoid long, unprotected paths inside the equipment.

3.15.5.4 Control of Apertures
Unnecessary apertures shall be avoided. Only those shield openings needed to achieve proper functioning and operation of the equipment shall be provided. Controls, switches, and fuse holders shall be mounted so close metal-to-metal contact is maintained between the cover housing of the devices and the case. Metal control shafts shall be grounded in accordance with paragraph 3.13.5.1. Where nonconductive control shafts are necessary, a waveguide-below-cutoff metal sleeve peripherally bonded to the case shall be provided for the shaft. The cutoff frequency for the circular waveguide shall be considerably higher than the equipment operating frequency. The length of the sleeve shall be no less than four times its diameter. Pilot lights shall be filtered or shielded as needed to maintain the required degree of shielding effectiveness. Ventilation and drainage holes shall not penetrate RF compartments if at all possible. If necessary, ventilation and drainage holes shall utilize waveguide-below-cutoff honeycomb or other appropriate screening. Care shall be taken to assure that honeycomb and screens are well bonded to the shield completely around the opening.

3.15.5.5 Bonding and Grounding of Compartment Shields
All shields shall be grounded. Bonding shall be accomplished in accordance with paragraph 3.14.

3.15.5.6 Wire and Cable Routing
The routing and layout of wires and cables shall be performed in a manner that does not

jeopardize the integrity of the equipment shield. High level signals shall be routed as far as feasible from low level signals. AC power cable and control lines subject to large transients shall be routed away from sensitive digital or other susceptible circuits. Shielded cables shall be used for either extremely low or high level signals. Cable shields shall be grounded in accordance with the requirements of paragraphs 3.15.3.2, 3.15.3.3 and 3.10.8.2.

3.15.5.7 Circuit Layout

Circuit layout techniques that provide maximum practical separation between high and low level signals shall be employed. High level pulse, switching, and power circuits shall be separated from low level digital, analog and similar sensitive circuits. Sensitive conductors internal to circuits shall not be laid out parallel to unshielded wires and cables subject to external interference. Shield terminations and connections to the signal reference ground shall be as short as practical.

3.16 Electrostatic Discharge (ESD) Minimization, Control and Prevention Requirements

Modern electronic and electronically controlled electrical equipment with high-speed, closely spaced circuitry and miniaturized components is highly susceptible to damage from Electrostatic Discharge (ESD). This phenomenon is most often related to operator contact with ESD susceptible items and must be considered when troubleshooting equipment. The requirements of this section are designed to reduce the frequency of ESD events and to minimize the effects. All electronic circuitry that contains miniaturized or solid-state components shall be considered ESD susceptible. Additional guidance on ESD is available in the FAA Orange Book on Electrostatic Discharge, dated July 1, 1996.

3.16.1 ESD Sensitivity Classification.

Classification of items as ESD sensitive shall be in accordance with the Human Body Model testing procedures and requirements of ESD-STM 5.1, Electrostatic Discharge (ESD) Sensitivity Testing, Human Body Model (HBM) – Component Level. Electronic parts, components, and assemblies shall be classified as either sensitive or supersensitive. Items that will fail from ESD at 1000 to 16000 volts shall be classified as ESD sensitive. Those items that will fail below 1000 volts shall be classified as supersensitive. Any exceptions to this guidance shall be through the OPI of this document. ESD sensitive devices with sensitivity of less than +/- 200 volts may require additional ESD protection measures than those specified in this standard. ESD susceptible items shall not be exposed to an electrostatic field (E-field) greater than 100 volts/meter or brought closer than 24" to known static generators or non-essential insulative materials.

3.16.2 ESD Protection Requirements.

All NAS electrical and electronic equipment, subassemblies, and components subject to damage from exposure to electrostatic fields or electrostatic discharge (ESD) shall be protected in accordance with the protection requirements herein. ESD controlled areas shall be provided for all operations, storage, repair, and maintenance spaces used for electrical and electronic equipment or subassemblies that are subject to damage from static electricity or ESD.

3.16.3 Circuit and Equipment Design.

3.16.3.1 Circuit Design and Layout.
The design, layout, and packaging of assemblies, circuits, and components integrated into electrical and electronic equipment shall incorporate methods and techniques to reduce susceptibility to ESD

3.16.3.2 Component Protection.
Integrated circuits, discrete components, and other parts without internal ESD protection that are inherently susceptible to ESD and have exposed conductive paths shall be externally protected by a capacitor, SAS, or a varistor. Protective components shall be installed as close as possible to the ESD susceptible item.

3.16.3.3 ESD Withstand Requirements.
In the installed and operational configuration, all equipment cabinets, enclosures, racks, controls, meters, displays, test points, interfaces, etc, shall withstand a static discharge of 15,000 volts per ESD Association Standard Test Method ESD-STM 5.1, Electrostatic Discharge Sensitivity Testing – Human Body Model (HBM). To successfully pass HBM testing requirements, equipment that is tested shall not suffer any operational upset or damage to any component or assembly.

3.16.4 Classification of Materials.

3.16.4.1 General
Most materials and products that are used to control and prevent ESD are made of conductive or static dissipative materials that are classified by their resistive properties. Antistatic materials are an exception to this and are classified by their propensity to generate static electricity from triboelectric charging. Any material used for construction of ESD protected areas (with the exception of antistatic materials) shall meet the resistive properties specified for type and use of the material. Materials that will tribocharge to greater than +/- 200 volts (EIA-625), if the material were to contact and separate from itself or from other materials, shall not be used in ESD controlled areas.

3.16.4.2 Conductive Materials.
Those materials with a surface resistance (R_{tt}) less than 1.0×10^5 ohms per ANSI/EOS/ESD S11.1 shall be considered to be conductive. Conductive ESD control materials may create an electrical shock hazard if used near energized equipment and shall not be used for ESD control worksurfaces, table top mats, floor mats, flooring, or carpeting where personnel may come in contact with energized electrical or electronic equipment. Conductive ESD control materials are not to be used in any other application where their use could result in EMI or RFI that would be created by rapid, high voltage ESD spark discharges. Any exceptions to this guidance shall be through the OPI of this document.

3.16.4.3 Electrostatic Shielding Materials.
Electrostatic shielding materials are a subset of conductive materials with a surface resistance equal to or less than 1.0×10^3 ohms (ANSI/EOS/ESD S 11.11). Electrostatic shielding materials

may be used as barriers for protection of ESD sensitive items from electrostatic fields where required.

3.16.4.4 Electromagnetic Shielding Materials.
Electromagnetic shielding materials with highly conductive surfaces (< 10 ohms) or specifically designed composite materials that absorb and reflect electromagnetic radiation over a broad range of frequencies may also be used as barriers where required to protect ESD sensitive items from electromagnetic fields.

3.16.4.5 Static Dissipative Materials.
Those materials with a surface resistance greater than 1.0×10^5 ohms but less than or equal to 1.0×10^{12} ohms (ANSI/EOS/ESD-S11.11) are classified as static dissipative materials. Static dissipative materials with a surface resistance less than or equal to 1.0×10^9 ohms shall be used to provide controlled bleed-off of accumulated static charges in ESD controlled areas. Static dissipative materials with a surface resistance of greater than 1.0×10^9 ohms but less than 1×10^{12} ohms shall not be used for applications where controlled bleed-off of accumulated static charges is essential. Any exception to this guidance shall be through the OPI of this document.

3.16.4.6 Antistatic Materials.
Any material that inhibits or has a low propensity to generate static electricity from triboelectric charging shall be considered antistatic. Antistatic ESD control items and materials used for construction of ESD controlled areas in new or renovated facilities shall not tribocharge to greater than +/- 200 volts when being used for their intended application. Antistatic materials with a surface resistance greater than 1×10^9 ohms shall not be used for ESD protective work surfaces, tabletop mats, floor mats, flooring, and carpeting when charge dissipation is the primary consideration. If the surface resistance (R_{tt}) of an antistatic material is greater than 10^{12} ohms it shall normally be considered to be too resistive for use in ESD controlled areas. Use of antistatic items and materials that utilize hygroscopic surfactants that depend on ambient humidity to promote absorption of water shall be limited. Only antistatic materials that are intrinsically antistatic and will retain their antistatic properties shall be used in ESD controlled areas. Exceptions to this guidance shall be through the OPI of this document. However, if controlled charge bleed-off is not a primary consideration, an antistatic material with a surface resistivity below 1.0×10^{12} ohms or volume resistivity less than 1.0×10^{11} ohm-cm can be used to inhibit triboelectric charging.

3.16.4.7 Static-Generative Materials, Non-Conductors, and Insulators
Materials having a surface resistance greater than 1.0×10^{12} ohms (ANSI/EOS/ESD-S 11.11) shall be considered to be insulators and a possible source of triboelectric charging. These include common plastics, Plexiglas, Styrofoam, Teflon, nylon, rubber, untreated polyethylene, and polyurethane. Their use shall be minimized where ESD sensitive items are located.

3.16.5   Protection of ESD Susceptible and Sensitive Items

3.16.5.1 Static Protected Zone
A static protected zone shall be a volume or area where unprotected ESD sensitive items will be

protected from direct contact with electrostatic potentials greater than +/- 200 volts, electrostatic fields greater than 100 volts/meter, or radiated electromagnetic interference (EMI) and radio frequency interference (RFI) produced by rapid, high voltage ESD spark discharges. Static protected zones shall be incorporated into the construction of ESD special protection areas, ESD protected workstations, and ESD protected storage areas.

3.16.5.2 ESD Special Protection Areas.

Special protection areas shall be designated areas that require extraordinary ESD control measures to accomplish the following: minimize triboelectric charging; control bleed-off and dissipation of accumulated static charges; neutralize charges; and minimize the effects of E-Fields, H-Fields, and EMI and RFI from ESD spark discharges. Areas within a facility that shall be designated as ESD special protection areas are: air traffic operations areas (e.g., tower cab, TRACON, ARTCC control rooms, AFSS, etc.); electronic equipment rooms; storage areas for ESD susceptible components, subassemblies, circuit cards, etc.; and areas that contain personal computers and Local Area Networks (LANs) that are connected to or interface directly with NAS electronic equipment. All other locations where jacks, plug in connectors or interfaces of ESD sensitive electronic equipment are exposed and vulnerable to damage from ESD by direct human contact shall also be designated as ESD special protection areas.

3.16.5.2.1   ESD Controls Required for ESD Special Protection Areas.

The following minimum ESD control measures shall be implemented in all areas designated as ESD special protection areas:

3.16.5.2.1.1   ESD Groundable Point (GP).

Each ESD control material, surface, or item used in an ESD controlled area shall have a designated groundable point (GP) to provide ease of connection to a common groundable point. The common groundable point (GP) shall be designated as ESD Common Point Ground or ESD ground. The common groundable point shall accommodate electrical connections from the groundable points of all elements of the ESD control system in the area served and shall also be connected to the nearest multipoint ground.

3.16.5.2.1.2   Grounded Static Dissipative Surfaces.

All work surfaces which includes work surface laminates, paints and sealers, writing surfaces, table tops, consoles, and table top mats shall be static dissipative and connected to a common ground point (e.g., ESD common point ground or ESD ground) or directly to the multipoint grounding system in the area served. The point-to-point resistance and surface to ground resistance of static dissipative work surfaces shall be greater than 1.0×10^6 ohms and less than 1.0×10^9 ohms (ESD-STM 4.1, for the Protection of Electrostatic Discharge Susceptible Items - Worksurfaces - Resistive Characterization).

3.16.5.2.1.3   Limiting the Use of Non-ESD Control Materials.

Materials that will tribocharge (e.g., generate electrostatic potentials by contact and separation with themselves or other materials) shall not be used for construction in ESD special protection areas. Insulative materials and any other non-essential triboelectric charge generators that generate potentials in excess of +/- 200 volts shall not be permitted within 24" of ESD special protection areas.

3.16.5.2.1.4   Static Dissipative Chairs.
Chairs (e.g., seating) provided for ESD special protection areas shall incorporate a continuous path between all chair elements (e.g., cushion and arm rests) to the ground point of greater than 1.0×10^4 ohms to less than 1.0×10^9 ohms. The ground point for ESD chairs shall be considered to be the static dissipative or conductive casters that provide electrical continuity from all elements of the chair to an ESD control floor (e.g., ESD control carpeting, tile, or floor mats) that is properly bonded to an appropriate ESD ground. ESD control chairs must be tested and meet the requirements of ESD Association Standard Test Method, ESD STM 12.1, Seating - Resistive Measurement.

3.16.5.2.1.5   Static Dissipative ESD Control Floor Coverings.
Static dissipative ESD control floor coverings shall include static dissipative tile, carpeting, static limiting floor finishes, and floor mats. Floor coverings in ESD special protection areas shall have a point-to-point resistance and surface-to-ground resistance of greater than 2.5×10^4 ohms and less than 1.0×10^9 ohms (ANSI/ESD-S7.1, Resistive Characterization of Materials - Floor Materials). These floor coverings shall be bonded to a ground point (e.g., ESD common point ground) or directly to the multipoint grounding system in the area served in accordance with paragraph 3.16.5.2.1.1.

3.16.5.2.1.6   Relative Humidity Control.
Relative humidity in ESD special protection areas shall be maintained within the range of 40 – 60%.

3.16.5.3   ESD Signs, Labels, Cautions, and Warnings.
ESD warning signs that include ESD sensitive device warning symbols with appropriate cautions and warnings shall be posted in ESD special protection areas and all other ESD controlled areas. Exterior cabinets of ESD sensitive electronic equipment shall also be marked or labeled with an ESD sensitive device symbol with a warning that is visible from at least 3 feet from the equipment.

3.16.5.4   ESD Protected Workstations.
In addition to the ESD controls specified in paragraph 3.16.3.2 through paragraph 3.16.3.3 the following shall apply to all ESD protected workstations.

3.16.5.4.1   ESD Protected Workstation Minimum Requirements
All ESD control items at an ESD protected workstations shall be connected to a common groundable point (e.g., ESD common point ground) that is connected to the multipoint grounding system in the area served. ESD protected workstations shall provide a means of personnel grounding (e.g., grounded wrist strap or conductive footwear in conjunction with ESD static dissipative floor or mat) and shall have a grounded static dissipative work surface, grounded static dissipative ESD control floor or mat, and be free of all non-essential static charge generators. Storage containers that may be provided at ESD protected workstation shall provide ESD protection and shall also be connected to the ESD ground. All outlets at ESD protected workstations shall be protected with ground fault circuit interruption (GFCI) capability to minimize danger to grounded personnel from electrical shock..

3.16.5.4.2 Use of Ionization.
Selective use of bench top or area ionizers may be considered at ESD protected workstations if static generative items (e.g., insulators) are deemed essential and cannot be removed from ESD protected workstation areas or if grounding of mobile personnel would be cumbersome or create a safety hazard.

3.16.5.4.3 Identification of ESD Protected Workstations.
The boundaries of all ESD protected workstations shall be clearly defined. The boundaries of ESD protected workstations shall extend a minimum of 24" beyond where ESD sensitive items will be located and should be marked with yellow tape. ESD warning signs that are yellow with black markings and lettering shall be posted that will be visible to anyone entering these areas. Signs shall include an ESD sensitive electronic device warning symbol and appropriate warnings and cautions.

3.16.5.5 ESD Protective Storage Areas.

3.16.5.5.1 Shelves, Bins, and Drawers.
Shelves, bins, and drawers shall be static dissipative and electrically continuous with the support structure of the storage shelves, bins, or container.

3.16.5.5.2 Grounding.
The storage container metal support structure shall have a groundable point (GP) that shall be connected to the ESD common point ground or directly to the nearest multipoint ground. The resistance from the ground point of storage containers, shelving, cabinets, and bins used to store ESD sensitive items to the multipoint ground system shall be less than one ohm.

3.16.5.5.3 Personnel Grounding.
Wrist straps shall be equipped with 1megaohm or greater series resistance to protect personnel. Standard 0.157" banana jacks for personnel grounding wrist straps shall be connected to an ESD ground point or directly to the multipoint ground of the area served. The resistance from a banana jack to a ground point and/or to the nearest multipoint ground shall be less than one ohm.

3.16.5.5.4 Prohibited Materials in ESD Protective Storage Areas.
Static generative (e.g., insulative) materials shall not be used for construction in any areas where ESD sensitive items will be stored. All materials that can generate potentials greater than +/- 200 volts shall be a minimum of 24" from ESD protected storage areas.

3.16.5.5.5 Resistance to ESD Ground for Shelves, Drawers, and Bins.
All surfaces and drawers of the storage media provided shall be made with static dissipative materials and meet the requirements and be tested the same as work surfaces (ANSI/EOS/ESD S 4.1). The surface-to-surface resistance (R_{tt}) and surface-to-ground resistance (R_{tg}) from the shelves, bins, and drawers of storage containers that will be used to store unprotected ESD sensitive items shall be greater than 1.0×10^6 ohms and less than 1.0×10^9 ohms (ESD Association Advisory, ESD ADV 53.1, for the Protection of Electrostatic Discharge Susceptible Items - ESD Protective Workstations).

3.16.5.5.6 Identification of ESD Protective Storage Areas.
The boundaries of all ESD protective storage areas shall be clearly defined. Boundaries of ESD protective storage areas shall extend a minimum of 24" beyond where ESD sensitive items will be located and should be marked with yellow tape. ESD warning signs that are yellow with black markings and lettering shall be posted that will be visible to anyone entering these areas. Signs shall include an ESD sensitive electronic device warning symbol and appropriate warnings and cautions.

3.16.6   Hard and Soft Grounds.

3.16.6.1 Hard Grounds.
Any item, material, or product that is a part of the ESD control system that is intentionally or unintentionally connected directly to an ESD ground point (e.g., ESD common point ground reference) or directly bonded to a multipoint ground shall be considered to be hard grounded. Unless specified otherwise or justified by the OPI for this document, ESD control worksurfaces, cabinets, flooring, carpeting, test equipment, and any other items used for ESD control shall be hard grounded to an ESD common point ground reference as specified above.

3.16.6.2 Soft Grounds.
A soft ground is the intentional connection to ground through a series resistor that will limit current flow to a predetermined maximum safe level if a person connected to the soft ground were to come into direct contact with a known potential. The soft ground concept shall only be used in personnel grounding skin contact devices such as wrist straps, leg or ankle straps, conductive shoes, and heel or toe grounders. All other elements of the ESD control system shall be hard grounded. Any exceptions to this guidance shall be through the OPI of this document. The nominal resistance of the resistor used for soft grounding of personnel shall be not less than 1.0×10^6 ohm unless otherwise specified by the OPI for this document.

3.16.7   ESD Control Floor and Coverings
All ESD control floors and floor coverings shall be installed, grounded, and initially tested only by trained installers. ESD control floors and floor coverings shall be bonded to the nearest multipoint ground at a minimum of four locations using 2" wide 26 gauge copper, copper foil, conductive fabric grounding ribbon, or stranded wire that makes electrical contact with the underside of the floor material or is embedded in the conductive permanent or releasable adhesive used to lay the floor. Refer to Table III for equivalent wire sizes.

3.16.7.1 Surface Resistivity (R_{tt}).

Surface resistivity (R_{tt} - Resistance top-to-top or surface-to-surface) of ESD control floors, carpets or floor mats shall be not be less than 2.5×10^4 ohms or more than 1.0×10^8 ohms (ANSI/ESD S 7.1). A minimum of 5 readings at different locations on the floor surface shall be taken and averaged together for each 500 square feet (or fraction thereof) of floor surface. These readings shall be recorded in the Facility Reference Data File.

3.16.7.2 Resistance Surface-to-Ground.
Resistivity from the floor surface to ground (R_{tg} - Resistance top-to-ground) of ESD control floors, carpets, or floor mats shall be greater than 2.5×10^4 ohms and less than 1.0×10^8 ohms (ANSI/ESD S 7.1). A minimum of 5 readings shall be taken at different locations on the floor surface and averaged together for each 500 square feet (or fraction thereof) of floor surface. These readings shall be recorded in the Facility Reference Data File.

3.16.7.3 Triboelectric Charging Limitation.
ESD control floors, carpets, or floor mats shall limit and control generation and accumulation of static charges to less than +/- 200 volts in ESD controlled areas.

3.16.7.4 Raised Floor Installation.

3.16.7.4.1 Resistance from Carpet Surface to Pedestal Understructure.
Carpet tiles that are installed on raised floor panels with conductive adhesive shall have a maximum resistance from the carpeted surface of the raised floor to the pedestal of not less than 2.5×10^4 nor more than 1×10^8 ohms.

3.16.7.4.2 Panel to Understructure Resistance.
Panel-to-understructure (metal-to-metal) contact resistances between individual access floor panels and the raised floor understructure shall be 10 ohms or less.

3.16.7.4.3 Carpet Tile Installation on Raised Floor Panels.
Individual carpet tiles may be installed on raised floor panels with either permanent or releasable conductive adhesive depending on the application.

3.16.7.4.3.1 Grounding.
There shall be a minimum of 4 connections from the carpeting undersurface and conductive adhesive to the raised floor panel understructure with 2" x 24" 26 AWG copper or copper foil strip per 1,000 square feet of installed ESD control carpeting.

3.16.7.4.4 Carpet Tile Installation with Positioning Buttons.
Carpet tiles may be installed with four permanently attached ultrasonically welded positioning buttons that can be inserted into positioning holes of specially designed access floor panels. The bare panel shall be coated with electrically conductive epoxy paint that will provide a maximum resistance of 1.0×10^3 ohms between the electrically conductive vinyl backing of the carpet tile to the raised floor understructure when the carpeting is mated with the access floor panel. These tiles using ultrasonically welded positioning buttons shall meet the surface to surface and surface to pedestal resistance requirements listed in paragraphs 3.16.7.1 and 3.16.7.2.

3.16.8 ESD Protective Worksurfaces.
Static dissipative materials or electrostatic dissipative laminates shall be used to cover all worksurfaces, consoles, workbenches, and writing surfaces in areas that contain ESD sensitive equipment and in all areas designated as ESD special protection areas, static-safe zones, and ESD protected areas.

3.16.8.1 Requirements for ESD Protective Worksurfaces.
Static dissipative worksurfaces shall be provided for new or upgrade facilities unless otherwise specified. Permanent static dissipative worksurfaces shall be connected to the closest ESD common point ground, element of the multipoint grounding system, or multipoint ground plate. Permanent ESD protective static dissipative worksurfaces shall have a resistance of not less than 1.0×10^6 ohms point-to-point (R_{tt}) in accordance with ANSI/EOS/ESD-S 4.1, ESD Protective Worksurfaces - Resistive Characterization. Permanent ESD protective worksurfaces shall have a resistance from their surface to the groundable point (R_{tg}) of not less than 1.0×10^6 ohms and not more than 1.0×10^9 ohms (ANSI/EOS/ESD-S 4.1).

3.16.8.1.1 Acceptable Worksurface Types.
ESD protective worksurfaces used in new and upgraded FAA facilities shall meet the requirements of MIL-PRF-87893, Performance Specification, Workstation, Electrostatic Discharge Control and MIL-W-87893, Military Specification, Workstation, Electrostatic Discharge (ESD) Control.

3.16.8.1.2 Type I Worksurface - Hard.
Type I worksurfaces shall be constructed of rigid static dissipative materials of any color having an average Shore D hardness in excess of 90. Two male or female 10 mm (0.395") ground snap (female) or stud (male) fasteners shall be installed on both corners on one of the longest sides of the worksurface to accommodate the male or female snap or stud fastener of the common point grounding cord. The locations of the two snaps or studs shall be 2" in from each corner.

3.16.8.1.3 Type II Worksurface - Cushioned.
Type II worksurfaces shall be constructed of cushioned static dissipative materials of any color having an average Shore A (ATSM D2240) hardness in excess of 45 and less than 85. Two male or female 10 mm (0.395") ground snap (female) or stud (male) fasteners shall be installed on both corners on one of the longest sides of the worksurface to accommodate the male or female snap or stud fastener of the common point grounding cord. The locations of the two male or female snaps or studs shall be 2" in from each corner. No low-density open-cell materials shall be used for Type II worksurfaces.

3.16.8.2 Static Dissipative Laminates.
High pressure, multi-layer static dissipative laminates shall be used to cover surfaces such as plywood, fiber board, particle board, bench tops, counter tops, and consoles in ESD controlled areas and special protection areas. The laminate shall include a buried conductive layer to provide for ease of grounding using a through bolted pressure type ESD grounding terminus. Static dissipative laminates shall be applied to surfaces using conventional contact adhesives.

3.16.8.3 Grounding of Laminated Surfaces.
The maximum resistance across the surface (R_{tt}) of the static dissipative laminate shall be less than 1.0×10^8 ohms and the maximum resistance from the surface of the laminate to ground (R_{tg}) shall be less than 1.0×10^7 ohms (ESD-STM 4.1). A minimum of 5 measurements of each shall be taken and averaged together. The values of the measurements and averages shall be recorded and maintained in the Facility Master File.

3.16.9 **Static Dissipative Coatings.**
Permanent clear or colored coatings (e.g., paint) used in ESD controlled areas shall dissipate static electricity. The surface resistance of static dissipative coatings shall not be less than 1.0×10^5 ohms or greater than 1.0×10^{10} ohms.

4. QUALITY ASSURANCE PROVISIONS

4.1 Electromagnetic Compatibility and Quality Assurance

4.1.1 General
A comprehensive plan for the application of various sections of this document is required to assure the compatible operation of equipment in complex systems. Additional considerations of this section shall be implemented to reduce susceptibility and emissions of equipment.

4.2 Requirements
The emission and susceptibility limits contained in MIL-STD-461 shall be applied unless otherwise specified. An EMI Control and Test Plan shall be developed in accordance with MIL-HDBK-237 to assure compliance with the applicable requirements. The plan shall include a verification matrix to track the satisfaction of requirement by test, analysis or inspection. Analyses or tests performed as part of paragraph 3.4 shall be identified in this plan.

4.3 Approval
Control Plans and Test Plans shall be submitted to the Contracting Officer for approval.

5. PREPARATION FOR DELIVERY
Section is not applicable to this standard

6. NOTES

6.1 Definitions

6.1.1 Access Well
A covered opening in the earth using concrete, clay pipe or other wall material to provide access to an EES (EES) connection.

6.1.2 Air Terminal
That component of a lightning protection system specifically designed to intercept lightning strikes.

6.1.3 Armored Cable
Power, signal, control or data cable having an overall armor constructed of ferrous (steel) material that provides both structural protection and electromagnetic shielding for direct buried cables.

6.1.4 Arrester
Components, devices or circuits used to attenuate, suppress, limit, and/or divert adverse electrical (surge and transient) energy to ground. The terms arrester, suppressor and protector are used interchangeably except that the term arrester is used herein for components, devices and circuits at the service disconnecting means.

6.1.5 Bond
The electrical connection between two metallic surfaces used to provide a low resistance path between them.

6.1.6 Bond, Direct
An electrical connection utilizing continuous metal-to-metal contact between the members being joined.

6.1.7 Bond, Indirect
An electrical connection employing an intermediate electrical conductor between the bonded members.

6.1.8 Bonding
The joining of metallic parts to form an electrically conductive path to assure electrical continuity and the capacity to conduct current imposed between the metallic parts.

6.1.9 Bonding Jumper
A conductor installed to assure electrical conductivity between metal parts required to be electrically connected.

6.1.10 Branch Circuit
The circuit conductors between the final overcurrent device protecting the circuit and the load served.

6.1.11 Brazing
A joining process using a filler metal with working temperature above 800°F but below the melting point of the base metal(s). The filler material is distributed by capillary action.

6.1.12 Building
The fixed or transportable structure which provides environmental protection.

6.1.13 Bulkhead Plate
A metallic plate located where conduits, conductor, waveguides etc first enter the facility. The bulkhead plate provides a convenient point for the grounding of conduits, conductors and waveguides entering the facility or structure.

6.1.14 Cabinet
An enclosure designed either for surface mounting or flush mounting and is provided with a frame, mat, or trim in which a swinging door or doors are or can be hung.

6.1.15 Cable
A fabricated assembly of one or more conductors in a single outer insulation. Types include axial, armored and shielded.

6.1.15.1 Cable, AC (not the same as armored (DEB) cable)
Type AC cable is a fabricated assembly of insulated conductors in a flexible metallic enclosure.

6.1.15.2 Cable, Armored Direct Earth Burial (DEB)
Cable with a ferrous shield designed to provide both physical and electromagnetic protection to the conductors.

6.1.15.3 Cable, Axial
Cable where all conductors are oriented on a single axis. Examples include coaxial, biaxial, and triaxial cables

6.1.15.4 Cable, Shielded
Cable with a metallicized or braid shield to improve resistance to electromagnetic interference (EMI).

6.1.16 Case
A protective housing for a unit or piece of electrical or electronic equipment.

6.1.17 Chassis
The metal structure that supports the electrical or electronic components which make up the unit or system.

6.1.18 Clamp Voltage
Clamp voltage is the voltage that appears across the SPD terminals when the suppressor is conducting a surge or transient current.

6.1.19 Conductor
Bare or insulated, see below.

6.1.19.1 Conductor, Bare
A conductor that has no covering or electrical insulation.

6.1.19.2 Conductor, Insulated
A conductor encased within material of composition and thickness recognized by the NEC as electrical insulation.

6.1.19.3 Conductor, Lightning Bonding (Secondary)
A conductor used to bond a metal object, within the zone of protection and subject to potential build up different from the lightning current, to the lightning protection system.

6.1.19.4 Conductor, Lightning Down
The down conductor serves as the path to the earth grounding system from the roof system of air terminals and roof conductors or from an overhead ground wire.

6.1.19.5 Conductor, Lightning Main
The main conductors are the conductors intended to carry lightning currents between air terminals and ground terminations. These can be the roof conductors interconnecting the air terminals on the roof, the conductor to connect a metal object on or above roof level that is subject to a direct lightning strike to the lightning protection system, or the down conductor.

6.1.19.6 Conductor, Lightning Roof
Roof conductors interconnecting all air terminals to form a two-way path to ground from the base of each air terminal.

6.1.20 Crowbar
In surge protective devices (SPD), the term "crowbar" refers to a method of shorting a surge current to ground in surge protective devices. This method provides protection against more massive surges than other types, but lowers the voltage below the operational voltage of the electronic equipment causing noise and operational problems. It also permits a follow-on current that can cause damage.

6.1.21 Current Issue
See Issue, Current

6.1.22 Earth Electrode System (EES)
A network of electrically interconnected rods, plates, mats, piping, incidental electrodes (metallic tanks, etc.) or grids installed below grade to establish a low resistance contact with earth.

6.1.23 Electromagnetic Interference (EMI)
Any emitted, radiated, conducted or induced voltage which degrades, obstructs, or interrupts the desired performance of electronic equipment.

6.1.24 Electronic Multipoint Ground System

An electrically continuous network consisting of interconnected ground plates, equipment racks, cabinets, conduit junction boxes, raceways, duct work, pipes, copper grid system, building steel, and other non-current-carrying metal elements. It includes conductors, jumpers and straps that connect individual items of electronic equipment to the electronic multipoint ground system.

6.1.25 Electronic Single Point Ground System

A signal reference network which provides a single point reference in the facility for equipment that requires single point grounding. The single point ground system is laid out in a manner that minimizes stray currents, and magnetic interference. It consists of conductors, plates and equipment terminals, all of which are isolated from any other grounding system except at the main ground plate.

6.1.26 Equipment Areas

Areas that contain electronic equipment used to support NAS operation. These include electronic equipment rooms, TELCO rooms, VORs, Radars etc.

6.1.27 Equipment Grounding Conductor

The conductor with the phase and neutral conductors used to connect non-current-carrying metal parts of equipment, raceways, and other enclosures to the system grounded conductor and/or to the grounding electrode conductors at the main service disconnecting means or at the point of origin (X_o bond) of a separately derived system.

6.1.28 Equipment

A general term including materials, fittings, devices, appliances, fixtures, apparatus, machines, etc, used as a part of, or in connection with, an electrical installation.

6.1.29 Facility Ground System

Consists of the complete ground system at a facility including the EES (EES), electronic multipoint ground system (MPG), electronic single point ground system (SPG), equipment grounding conductors, grounding electrode conductor(s), and lightning protection system.

6.1.30 Faraday Cage

A closed conducting surface, such as wire mesh, completely surrounding an object or person so as to protect from impinging electromagnetic waves. Unbonded penetrations of the cage will result in a near complete loss of effectiveness. Effectiveness of a Faraday Cage is dependent upon the conductivity and the spacing (aperture size) of the conducting elements.

6.1.31 Feeder

All circuit conductors between the service equipment or the source of a separately derived system and the final branch circuit overcurrent device.

6.1.32 Fitting, High Compression

See "Pressure Connector".

6.1.33   Ground
A conducting connection, whether intentional or accidental, between an electrical circuit or equipment and the earth, or to some conducting body that serves in place of the earth.

6.1.34   Grounded
Connected to earth or to some conducting body that serves in place of the earth.

6.1.35   Grounded Conductor
A system or circuit conductor that is intentionally grounded at the service disconnecting means or at the source of a separately derived system. This grounded conductor is the neutral conductor for the power system.

6.1.36   Grounded, Effectively
Permanently connected to earth through a ground connection of sufficiently low impedance and having sufficient current carrying capacity so that ground fault current which may occur cannot build up to voltages dangerous to personnel.

6.1.37   Grounding Conductor
A conductor used to connect equipment or the grounded circuit of a wiring system to the grounding electrode system. (In this standard, grounding conductors not related to or not used as part of NEC required electrical system grounding, are used for the electronic equipment grounding system).

6.1.38   Grounding Electrode
Copper rod, plate or wire embedded in the ground for the specific purpose of dissipating electric energy to the earth.

6.1.39   Grounding Electrode Conductor
The conductor used to connect the grounding electrode to the equipment grounding conductor and/or to the grounded (neutral) conductor of the facility at the service disconnecting means or at the source of a separately derived system.

6.1.40   High frequency
All electrical signals at frequencies greater than 100 kilohertz (kHz). Pulse and digital signals with rise and fall times of less than 10 μs are classified as high frequency signals.

6.1.41   Issue, Current
Current Issue (applied to versions of referenced documents) The current issue is the issue in effect on the date of a contract signing when that contract references this document. In all other cases it is the version of the referenced document in effect when planning for a design is accomplished.

6.1.42   Landline
Any conductor, line or cable installed externally above or below grade to interconnect electronic equipment in different facility structures or to interconnect externally mounted electronic equipment.

6.1.43 Line Replaceable Unit
Hardware elements whose design enables removal, replacement and checkout by organizational maintenance.

6.1.44 Low Frequency
Includes all voltages and currents, whether signal, control, or power, from up to and including 100 kHz. Pulse and digital signals with rise and fall times of 10 μs or greater are considered to be low frequency signals.

6.1.45 National Electrical Code (NEC) (NFPA-70)
A standard containing provisions considered necessary for safety governing the use of electrical wire, cable, equipment and fixtures installed in buildings. It is sponsored by the National Fire Protection Association (NFPA) under the auspices of the American National Standards Institute (ANSI).

6.1.46 Operational Areas
Areas used to provide NAS support such as IFR rooms, ARTCC control rooms, ATCT tower cabs and operations control centers.

6.1.47 OPI
OPI is an acronym for Office of Primary Interest. The OPI is assigned responsibility for maintaining and interpreting the document. When in doubt about the meaning of any requirements in this document contact the OPI for clarification and additional guidance.

6.1.48 Overshoot Voltage
The fast rising voltage that appears across transient suppressor terminals before the suppressor turns on (conducts current) and clamps the input voltage to a specified level.

6.1.49 Pressure Connector
For purpose of this document, "FAA approved pressure connectors" shall be those that use hydraulic crimpers to effect closure.

6.1.50 Rack
A frame in which one or more equipment units are mounted.

6.1.51 Reference Plane or Point, Electronic Signal (Signal Ground)
The conductive terminal, wire, bus, plane, or network which serves as the relative zero potential for all associated electronic signals.

6.1.52 Reverse Standoff or Maximum Continuous Operating Voltage (MCOV)
The maximum voltage that can be applied across transient suppressor terminals with the transient suppressor remaining in a non-conducting state.

6.1.53 Shield
A housing, screen, or cover which substantially reduces the coupling of electric and magnetic

fields into or out of circuits or prevents accidental contact of objects or persons with parts or components operating at hazardous voltage levels.

6.1.54 Structure
Any fixed or transportable building, shelter, tower, or mast that is intended to house electrical or electronic equipment or otherwise support or function as an integral element of the air traffic control system.

6.1.55 Surge
An over voltage of short duration occurring on a power line. Lightning or switching events may cause surges.

6.1.56 Susceptibility Level
The electronic equipment susceptibility level is the least of the damage, degradation, or upset levels considering all electronic components potentially affected by conducted or radiated transients.

6.1.57 Transient
An overvoltage or overcurrent pulse on a signal, control, or data line. Transients are typically a lightning related phenomenon.

6.1.58 Transient Suppressor
Components, devices or circuits designed for the purpose of attenuating and suppressing conducted transient and surge energy to ground to protect facility equipment. At the service disconnecting means, the term "arrester" is generally used instead of transient suppressor. The term surge protective device (SPD) is preferred.

6.1.59 Turn-on Voltage
The voltage required across transient suppressor terminals to cause the suppressor to conduct current.

6.1.60 Zone of Protection
The zone of protection is that space adjacent to a lightning protection system that has a reduced probability of receiving a direct lightning strike.

6.2 Acronyms and Abbreviations

The following are acronyms and abbreviations used in this standard

A	Amperes	L-L	Line to Line
AC	Alternating current	L-N	Line to Neutral
ANSI	American National Standards Institute	LRU	Line replacement unit
AWG	American Wire Gauge	m	Meter
Cm	Centimeter(s)	mA	Milliampere
Cmil	Circular mils	MCM	See kcmil
DC	Direct current	MCOV	Maximum continuous operating voltage
e.g.	For example	MHz	Megahertz
EES	Earth electrode system	MPG	Electronic multipoint ground system
EMI	Electromagnetic interference	mm	Millimeter(s)
EPP	Equipotential plane	NAS	National Airspace System
EOS	Electrical overstress	NEC	National Electrical Code
ESD	Electrostatic discharge	NEMA	National Electrical Manufacturers Association
Et.al.	And others	NFPA	National Fire Protection Association
FAA	Federal Aviation Administration	No.	Number
ft.	Foot (feet)	OPI	Office of Primary Interest
GP	Groundable point	PVC	Polyvinyl chloride
Hz	Hertz	RF	Radio frequency
i.e.	That is	RGS	Rigid galvanized steel
IEC	International Electrotechnical Committee	RFI	Radio frequency interference
in.	Inch(es)	RMM	Remote maintenance monitoring
IEEE	Institute of Electrical and Electronics Engineers	SAS	Silicon avalanche diode suppressors
kA	Kiloampere	SPD	Surge protective device
kcmil	Thousand circular mils	SPG	Electronic single point ground system
kg	Kilogram	SRG	Signal reference grid
kHz	Kilohertz	UL	Underwriters Laboratories
LAN	Local area network	µs	Microseconds
"	Inch(es)	V	Volts
#	Number	'	Foot(feet)
L-G	Line to Ground		

6.3 Guidelines

Engineering design guidelines are provided for lightning protection, grounding, bonding, shielding, and transient protection in FAA Orders 6950.19 and 6950.20. Guidance for EMI protection is in MIL-HDBK-253, and for electrostatic discharge (ESD) in NFPA 77, DOD-HDBK-263 and DOD-STD-1686.

www.ingramcontent.com/pod-product-compliance
Lightning Source LLC
Chambersburg PA
CBHW081826170526
45167CB00007B/2734